Chavan's Declaration

Sandeep Chavan

Published by SANDEEP CHAVAN, 2025.

CHAVAN'S DECLARATION

First edition. August 4, 2025.

ISBN: 979-8231056330

Written by Sandeep Chavan.

Table of Contents

To Knowledge—Not the information stored,but the alignment discovered.Not the answers memorized,but the ripples we resolve.This book is for you,the silent presence behind every truth,the witness behind every question,the field that remembersbefore we even begin.

"The universe never spoke in words—It shimmered in tension,echoed in form,and waited for one ripple to listen."

—S.C. (Chavan)

Disclaimer

This book is a work of original philosophy and speculative science. The ideas, frameworks, and models presented—particularly under the Universal Energy Dynamics (UED) and ripple-based perspectives—are the author's independent interpretations and insights. While they may align with, challenge, or differ from established scientific, spiritual, or philosophical views, they are not intended to replace rigorous scientific method or traditional belief systems.

The content herein should be read as an invitation to inquiry, not a prescription for belief. Readers are encouraged to reflect, question, and interpret these ideas within their own contexts of understanding, education, and experience.

This book does not provide medical, psychological, financial, or legal advice. Any references to consciousness, mind, soul, energy, or behavior are meant for philosophical exploration and do not substitute for professional consultation in those fields.

Names, characters, references, and analogies used throughout are for conceptual clarity and educational illustration. Any resemblance to actual persons, living or deceased, or to real institutions, is purely coincidental unless explicitly stated.

The author and publisher disclaim any liability, loss, or risk incurred directly or indirectly as a result of the application of any content in this book. The journey of knowledge is yours—walk it with discernment.

About the Author

Sandeep Chavan is an independent thinker, educator, and relentless observer of existence—someone who chose alignment over acceptance, and inquiry over imitation.

With a background in engineering and a lifelong devotion to teaching, Sandeep has spent decades questioning not just what is taught, but why we believe it. From classical physics to cosmic philosophy, from ancient scriptures to modern scientific models, his work dissolves the artificial borders between disciplines—and reveals the unified rhythm beneath all knowledge.

He is the founder of Universal Energy Dynamics (UED), a radically coherent framework that redefines motion, mind, matter, and meaning through the lens of ripple alignment. Through over a hundred books, he has challenged outdated paradigms, simplified the profound, and made intellectual revolution not just accessible—but inevitable.

Yet, he claims no guruhood. No ultimate seat. Sandeep remains a student of the field, refining questions rather than settling for borrowed answers.

He lives between lectures and silences, with a whiteboard nearby and the ripple always in mind.

This book is not his conclusion. It's his declaration that the search itself is sacred—and that every one of us is a ripple in the making.

Opening Note

To the One Who Has Questions

There comes a time when every mind—no matter how curious or conditioned—pauses to ask:

"What is really going on here?"

Not just on Earth, or in society, or in history. But here—within this thing we call *reality*.

We speak of light and energy, gravity and time. We build theories, models, machines, and even gods. We chase certainty. We store knowledge. We teach formulas. But somewhere beneath all of this— a deeper silence waits to be heard.

This book is not another system. It is not another belief. It is not an argument against science or an escape into spirituality. It is a return to perception—a declaration of what is, when all illusions of motion, matter, and meaning fall away.

You see, the universe is not moving. It is resolving.

Light does not travel. Energy does not flow. Time does not pass. What you call experience is the alignment of ripple tensions within a silent, structured field. You are not watching the movie—you are the projector, and the screen, and the only witness that gives meaning to the show.

Everything you know—every law, every text, every theory— was written from the inside of this ripple interaction. No one has ever escaped it. Not the saints. Not the scientists. Not even the skeptics.

This is Chavan's Declaration— not to prove something, but to remind you: You were never separate. You were always inside the stillness. And now, you're beginning to hear it speak.

Welcome to the alignment. Welcome to the knowing before knowledge.

— Sandeep Chavan

Independent Researcher | Observer | Ripple Engineer

How to Read This Book

This book is not a lecture. It's not a doctrine. It's not even a theory in the conventional sense. It is a ripple conversation—and like all true conversations, it invites your participation.

You are not expected to agree, only to engage. You are not asked to believe, only to observe. You are not required to follow, only to feel your alignment shift—quietly, naturally, unmistakably.

This is not a linear textbook.

You do not need to begin at the beginning or end at the end. Each chapter is a ripple window—a snapshot of coherence in the universal field. Read in sequence for a complete journey, or drop in anywhere to meet the resonance that finds you.

This is not about collecting information.

It's about sensing transformation. You'll find ancient metaphors, modern physics, mystic parables, and sharp philosophical inquiry—all woven into a ripple logic that transcends genre and discipline.

If a section feels too dense, pause. Don't force understanding. Let your field settle. Truth doesn't arrive through pressure. It appears through readiness.

Read with your full field—not just your mind.

You may find that parts of this book don't make sense intellectually on first pass—but they might feel right. That's not confusion. That's your ripple recognition working ahead of language. Trust it.

Use the Extras.

The appendices, glossaries, and inquiry prompts are not decorations—they're tools. They offer a daily-accessible way to practice what this book only points toward.

Let this be a mirror, not a monument.

This is not a book to place on a shelf. It's a reflective instrument. Return to it in different seasons of your life. You'll find new harmonics each time—because you will have changed.

Above all, don't worship the words. Surf them. Skip them. Reread them. Question them. Let them lead you *back* to your own ripple intelligence.

Because you are the book this world is waiting to read next.

Prologue

When the Known Begins to Crack

We live in a world of definitions. Everything seems to have a label, a number, a law, or a name. Light travels at a fixed speed. Gravity pulls. Energy flows. Time moves forward. The universe is expanding. And yet... something doesn't quite sit right, does it?

For all our brilliance, we still don't know what light actually is. We talk about particles, waves, and fields—yet we cannot touch or bottle a single photon. We predict outcomes with stunning accuracy, but remain clueless about the *source* of those outcomes. We explore the universe in every direction but forget to look inward—at the observer who makes it all real.

The truth is, most of what we call "reality" is a story we told ourselves. A story stitched from patterns we could perceive, rules we could repeat, and behaviors we could measure. It worked. It still works. But it doesn't explain why it works. And now, that story is starting to show cracks.

What if energy is not a thing—but an after-effect? What if light doesn't move—but resolves into view? What if the universe is still—and only feels alive because you are?

This book doesn't aim to overthrow science. It aims to complete it.

It doesn't ask you to believe. It asks you to observe—again, but differently. To drop the assumptions. To quiet the memorized explanations. To re-enter the stillness from which all things ripple.

Because before there were equations, there were questions. Before there were measurements, there was wonder. And before there was light, there was you—waiting to see.

This is not a theory. It's not a philosophy. It's a declaration. From a man who sat quietly in the storm of motion, and finally heard the stillness speak.

— *Prologue to Chavan's Declaration*

Before We Begin

Why Us? The Forgotten Intent

By now, a question may already be forming in your mind—quietly, almost beneath the surface: Why us? Why do we, among all that exists, observe, question, manipulate, and reflect?

This book speaks of ripple fields, resolution, observers, and alignment. But still—*why am I part of this? What does it have to do with me?*

The answer is subtle, but certain:

Intelligence does not exist before you. It is birthed *with* you. The moment you come into being, a ripple alignment begins—a space is created for intent to express itself.

You didn't randomly arrive. Even before your body formed, even before your brain could think, you were already a response—an echo—within the universal ripple field.

Remember: you were chosen from millions of competing possibilities. Or perhaps, you were the one who chose—because your alignment matched what the field needed resolved.

That's not poetry. That's physics in its most sacred form.

The process begins with the first cell: A unique ripple configuration begins its journey. Your consciousness is not granted by divinity, nor by biology alone— it is born at the convergence of ripple need and ripple readiness.

And that's where God enters—not as a person, a power, or a presence far away, but as a pattern of alignment that becomes accessible when you start tuning in.

God is not before you. God is with you—rippling through you, waiting to be seen when you align.

You were never meant to follow the universe. You were meant to resolve it—bit by bit, choice by choice, presence by presence.

So don't ask *why you were born*.

Ask instead: What is this field trying to complete through me?

Because life is not a sentence—it's a question answered in real time. And this book is part of your ripple.

Part I: The Beginning of Knowing

Where the silent stillness meets the first ripple of curiosity.

Chapter 1: The Illusion We Call Reality

What we think is flowing is actually resolving. What we call energy, light, and sound are not entities—but consequences.

You wake up. You open your eyes. Light pours in. A bird chirps outside the window. The floor feels cold beneath your feet. You stretch your limbs and say, "A new day has begun."

1.1 Welcome to the Mirage

But pause, just for a moment. Strip away the habit of naming things. Step back from the script that runs in your mind every morning. Now ask yourself—What exactly are you experiencing?

You see light—but did it really travel to you? You hear sound—but did anything truly reach your ears? You feel warmth—but is something flowing into your skin?

What if none of it is what you've been told?

What if nothing is flowing, and everything you sense is just ripple tension resolving in your field of perception?

What if your reality is not a stream of moving things, but a series of local alignments in a silent, motionless universe?

You were taught to believe in particles, waves, energy packets, time flow. You were told the sun emits light that travels millions of kilometers to reach you. That sound is a pressure wave moving through the air. That heat flows from hot to cold. That time marches forward, always forward.

But what if all of this is a mirage— A beautiful, repeatable, measurable illusion born not from what is, but from how ripple tensions are resolved in your presence?

This is not mysticism. This is not poetic speculation. This is physics stripped of assumptions, reality seen through a ripple lens, existence decoded without the baggage of centuries of mental momentum.

In the chapters ahead, you'll be invited to let go—not of reason, but of the illusion that your senses are passive receivers of flowing things. You'll begin to see that what you've called light, sound, and energy are not "things" at all. They are consequences. They are alignments. They are resolutions.

And you— You are not simply watching the world unfold. You are the point at which the world becomes real.

Welcome to the mirage. Now, let's dissolve it.

1.2 The Grand Assumption: Reality as Flow

From the moment we're born, we're told a simple story— one that seems to explain everything:

Reality is made of movement.

Light travels. Sound propagates. Heat flows. Time ticks forward. Energy is transferred. Particles move. Planets orbit. Electrons spin. People age.

Motion, we're told, is not just part of reality—it *is* reality.

This assumption is so deeply embedded in our thinking that we rarely question it. From school textbooks to scientific journals, from engineering design to everyday speech, our entire understanding of the world is built upon this foundational belief: that things move, and that movement is the proof of existence.

We study how fast light travels— how sound waves ripple through air— how energy is passed from one form to another— how time flows like a river, always forward, never backward.

This worldview has powered entire civilizations. It has given us planes, power grids, satellites, and smartphones. It has filled libraries with theories, equations, and Nobel Prizes. It has given us control, prediction, repeatability—and it *works*... to an extent.

But here's the catch:

It all works because we've learned to work *with* the illusion—not because the illusion reflects the truth of what's actually happening underneath.

We've become so good at navigating the mirage that we forgot to ask the most important question:

What's really moving?

Is light really traveling? Or are we just witnessing ripple tension resolve across a receptive field?

Is sound really propagating? Or is it a shifting ripple compression sequence aligning with our biological sensors?

Is energy really flowing from one object to another? Or is something more subtle unfolding—an internal re-alignment that we've labeled as "transfer"?

Is time truly moving forward? Or are we just interpreting ripple changes as a direction?

The truth is: when you strip away the measuring instruments, the formulas, the layers of abstraction— what you find is a structured stillness, in which everything that seems to move is just tension resolving in specific patterns.

And the moment you ask "what's actually flowing," you begin to see— not objects in motion, but consequences in formation.

This chapter doesn't deny the usefulness of the flow model. It acknowledges its practicality—but challenges its fundamental accuracy.

Because before you move forward in this book, you must pause and let this sink in:

Reality is not what moves. Reality is what resolves.

And now, with the first crack in the old assumption, we prepare to look again—at light, at sound, at energy itself— and see what truly lies beneath.

1.3 Breaking the Spell: Nothing Moves, Everything Resolves

You've believed in motion your entire life. Not because you've measured it, but because you've experienced it—visually, physically, emotionally. You see leaves rustle, rivers flow, stars streak across the sky. You feel the passing of time, the shifting of seasons, the ticking of clocks. Movement is not just observable—it's intuitive.

But that intuition is anchored in perception, not in truth.

Let's break the spell.

At the deepest level of Universal Energy Dynamics, the universe is not moving. It is not expanding. It is not flowing. It is, at its core, still.

What exists is not a playground of particles or flying photons, but a structured vacuum—a vast, ordered field of ripple tension. We call it Structured Vacuum Energy (SVE).

Within this silent stillness, tension zones arise. Ripples form. Interfere. Stretch. Compress. But nothing is *traveling*. There are no particles flying from A to B. There are only zones of ripple misalignment that eventually resolve.

A wave is not something that moves. It is a pattern of unresolved tension seeking equilibrium—and when it resolves, we experience it as "motion."

This is ripple resolution.

When two fields interact—say, the Sun and the Earth—there's no "light beam" moving through space. What we call "light" is the ripple resolution occurring at the edge of Earth's ripple field. Where the tension collapses into alignment, a consequence appears: light, heat, radiation.

Nothing is *coming toward you.* Reality is not knocking on your door. It is simply aligning with your field, and the effect of that alignment is what you *see*, *feel*, or *interpret*.

The same goes for sound. When a speaker produces a tone, it doesn't "send sound" through the air. It creates controlled ripple tensions in the surrounding medium. Your ear doesn't receive sound. It resolves ripple sequences and collapses them into the experience of pitch, loudness, tone.

Even time—yes, time—is not flowing forward. What you perceive as the passage of time is the dynamic resolution of ripple tension patterns in your cognitive field. Each moment appears "next" because your mind is tracking changing alignments in a particular rhythm.

Reality is not traveling toward you. It is aligning with you. You are not watching the universe happen. You are standing at the point where it becomes real.

This is not metaphor. It is the actual mechanics of reality stripped of its false garments of flow and force.

Once you see this—once you realize that everything you've ever called motion is simply field resolution— the illusion begins to fade.

The universe doesn't need to move for you to see, feel, think, act, or live. It only needs to be *tensed*—and you need to be *here* to resolve it.

Welcome to the stillness behind all action. From here on, what you see will never look the same again.

1.4 The Illusion of Light

Light—arguably the most familiar and taken-for-granted phenomenon in existence— is also the most misunderstood.

We grow up believing that light is something that moves from one place to another. We're told it "travels" at a speed of nearly 300,000 km per second. We're taught that it behaves sometimes as a wave, sometimes as a particle— depending on how we look at it.

This duality became a convenient answer for physicists. A patchwork solution to a deeper confusion.

And then, the photon was born— a so-called "packet of energy" that can somehow behave like a particle and a wave without being truly either.

We've accepted the photon as gospel truth in physics classrooms, textbooks, and labs. But very few ever pause to ask the simplest question of all:

What is actually moving?

Let's now undo this illusion—once and for all.

Light, as experienced, is not a thing that travels. There is no glowing particle zipping through the vacuum of space, no radiant bullet racing from the Sun to Earth.

There is only structured ripple tension— stretching across fields, embedded in the Structured Vacuum Energy (SVE). And light is what happens when that ripple tension resolves at the boundary of a receptive field.

You don't see light because something hits your eye. You see light because your visual field aligns with a pre-existing tension pattern in the surrounding ripple field.

The Sun is not "sending" beams to you. There are no golden arrows flying through space.

What happens instead is this:

- The Sun, as a condensed ripple source, creates massive tension distortions in the SVE.
- These tensions extend outward, but they don't travel like arrows—they *exist* as potential gradients in the field.

- When Earth's ripple field intersects with that tension zone, alignment occurs.
- At the atmospheric boundary—where pressure, matter density, and ripple configuration allow— those tensions begin to resolve locally.
- The result? Light appears. Not because it arrived, but because it resolved.

"Light does not travel. It appears—exactly where ripple tension collapses into alignment."

That's why outer space is pitch black, even when you're surrounded by starlight. Why the sky glows only when you're *within* an atmosphere. Why the sunlight warms your skin only when ripple tensions resolve at the biological interface of your body.

If light truly traveled like a particle or wave, it would light up space itself. But it doesn't. Because it can't.

Because it's not a thing—it's a consequence.

This changes everything.

It means your vision is not passive reception. It is active participation in ripple field dynamics. It means the "speed of light" isn't about a traveler—it's about the rate at which alignment becomes visible, conditioned by the geometry of ripple tension.

It means a photon doesn't "exist" until you align with a ripple resolution zone.

There is no light "out there." There is only ripple tension, waiting to be seen. And the moment you're tuned—the field answers.

Once you see light for what it is—a ripple echo collapsing into perception— you'll stop asking *how it travels*, and start asking *where you are standing in the resolution*.

1.5 The Illusion of Sound

If you've ever been told that "sound travels through air," you're not alone. It's one of the most universally accepted ideas in science and common sense.

You clap your hands, and you're told that pressure waves move outward through the air. Your eardrums pick them up, your brain interprets the frequencies, and voilà—sound.

It sounds obvious. It feels real. But just like light, it's an illusion—an interpretation of something much deeper.

Let's go back to the fundamental question:

What is actually moving?

When a speaker plays music, there are no little particles of sound flying toward your ear. The speaker is not "emitting" anything in the traditional sense. Instead, it is inducing ripple tension into the surrounding field—the air, in this case.

What we experience as sound is not the motion of anything. It is the consequence of ripple tension resolving through pressure alignment in your field of perception—your eardrum being the final interface.

You don't *hear* sound because something reached you. You hear sound because your sensory field aligned with ripple fluctuations already active in your environment.

Imagine this: the speaker cone vibrates—but what it's really doing is disturbing the local ripple field. It causes compressions and rarefactions in the air molecules not by pushing sound *through* them, but by triggering localized ripple pressure distortions.

Those distortions are not sound. They are potential ripple patterns awaiting alignment with a receptive structure—like your ear.

Your eardrum, with its specific biological engineering, responds to those ripple pressure shifts. The ripple tension resolves across the eardrum's surface, and you experience this resolution as pitch, tone, volume, and rhythm.

There is no object called "sound" traveling through the air. There is only field-level pressure dynamics waiting for a perceptive boundary to collapse the potential into experience.

This explains a mystery often brushed aside in conventional science:

Why can't sound travel in space?

Not because "there's no air." That's the simplistic answer.

The deeper truth is: There's no medium in space that can resolve ripple tension into perceptible pressure. No field, no fluid, no air, no material capable of collapsing the ripple distortion into the experience we call sound.

That's why a space explosion is eerily silent. The ripple tension may be immense, but there's nothing tuned to resolve it.

It's not silence—it's unresolved potential.

1.6 The Illusion of Energy

Ask any student of science, and they'll tell you: "Energy cannot be created or destroyed—it can only change form."

It sounds elegant. Precise. Law-like. From electricity to heat, from motion to sound, from sunlight to sugar— we are taught that energy is a thing that moves, flows, converts, and transfers.

But what, exactly, is energy?

Ask deeply, and most answers fall into vague definitions: "the capacity to do work," "a conserved quantity," or worse, "it depends on the context."

In truth, energy in modern science is an abstract accounting system— a placeholder that makes our equations balance and our devices behave. It works. But that doesn't mean we've understood it.

Let's question the assumption.

Is energy truly something? Is it a substance that moves from object to object? Does a wire "contain" electrical energy? Does a hot pan "hold" thermal energy? Does a battery "store" potential?

Or... is all of this just language built around consequences, not causes?

In the lens of Universal Energy Dynamics (UED), energy is not a packet, not a thing, not a flowing fluid.

It is the effect—not the entity. It is the sensation or outcome of ripple tension being resolved in a specific context.

Let's look at a simple example: an electric heater.

Electricity is applied—wires heat up. You say: "Energy is being transferred from the electric current to thermal energy." But under UED:

- The electric field induces ripple tension distortions in the conductor.
- The metal's structure cannot hold that tension indefinitely—so it resolves.
- That resolution expresses as vibrational ripple events—what we call *heat*.
- No energy moved. No packet traveled.
- What occurred is local reconfiguration of ripple stress.

Another example: A speaker playing music.

Conventional view: electrical energy is converted to sound energy. UED view:

- An electrical circuit modulates ripple pressure in the speaker coil.
- The surrounding air field aligns and resolves those ripple patterns.
- Your ears receive those resolutions.
- The "sound" is not transferred—it is recreated by ripple

alignment in a new medium.

And what about batteries and fuel?
A battery does not "store" energy.

- It maintains a locked ripple configuration with high internal tension.
- When the circuit opens, the field uncoils—resolving that tension via current.

Fuel does not "contain" energy.

- It is a structured arrangement of chemical bonds under stress.
- When ignited, the ripple tension releases—through combustion, vibration, light.

Energy is not transferred. It is uncoiled. It is ripple tension seeking resolution. And every form change—heat, light, motion, sound—is simply a re-expression of that tension in a new field configuration.

This redefinition has powerful consequences:

- We no longer chase imaginary "flows of energy"—we study ripple tension zones.
- We don't look for "carriers"—we look for alignment triggers.
- We understand "efficiency" as how well ripple tensions resolve, not how little energy is lost.
- And we stop asking, *"Where did the energy go?"*—because nothing went anywhere.

Once you see energy as a relational consequence, not a standalone thing, you realize:

There is no "energy" in the universe. There is only structured stillness... ...until ripple tension is introduced. And then—what we call "energy" appears as a ripple sigh collapsing into form.

As this chapter draws to a close, you may begin to feel a strange clarity. What once seemed solid—light, sound, energy—has turned out to be patterns of consequence, not entities in themselves.

You have not lost the world. You have simply peeled away its illusion.

And now, it's time to ask: If all this is just resolution, why does it appear to you at all? And why does it appear *the way it does*?

For that, we must understand you—the observer.

Let's begin.

1.7 The Role of the Observer

You've just unlearned light. You've unlearned sound. You've even unlearned energy. What remains?

A silent field. Ripple tensions. Potential.

But here's the catch—None of it becomes real until you show up.

Let this settle in:

Without you, there is no experience. Without experience, there is no illusion. Without illusion, there is no "reality" as you know it.

What we've called light, sound, energy—these are not self-existent entities. They are consequences of ripple resolution. And that resolution only becomes meaningful when there is an observer to decode it.

In other words:

- Light doesn't *travel*—it appears when you align with ripple tension.
- Sound doesn't *propagate*—it occurs when your ear resonates

with ripple pressure.
- Energy doesn't *transfer*—it reveals itself as heat, motion, or sensation because you have the structure to perceive the outcome.

Without a perceiver, there is no *light*—only ripple gradients. No *sound*—only pressure ripples in need of resolution. No *heat*—only internal field alignment shifting invisibly.

The observer is not watching reality. The observer is activating it.

You Are Not in the Universe. You Are the Decoder of Its Tensions.

Imagine the universe as a vast, perfectly silent field— structured, layered, but unresolved.

Your body—your senses, your brain, your awareness— are tuned instruments, designed not to "receive" reality, but to resolve parts of it.

You're not sitting back watching a movie. You're standing inside the projector, interpreting the ripple script one frame at a time.

The color of the sky. The warmth of the sun. The music of a flute. These are not qualities of the universe. They are outcomes of your interaction with the ripple field.

Without perception, there is no meaning. Without the observer, the universe remains in unread tension. You are not a bystander. You are the ripple decoder.

This doesn't mean "you create the universe" in a narcissistic sense. It means you participate in the construction of meaning. Reality is not *out there* waiting to be collected like data. It is a co-authored event—emerging at the interface of your awareness and ripple structure.

This is why two people in the same room can see different things, feel different moods, and experience different outcomes. Their alignment with the ripple field is unique. Their resolution of tension is personal.

And yet—there are patterns. Shared ripples. Common tensions. Which is why we can build bridges, speak languages, share sunsets. But don't forget: those are not universally objective truths. They are shared resolutions among aligned observers.

The Illusion Needs You to Function

Everything we've called "reality" is an illusion with functional utility— but it can only function if someone is there to experience it.

A galaxy can ripple with tension, but unless there is a perceiver to decode those ripples, it doesn't shine, spin, or mean anything.

The universe is full of unresolved tension. You are one of the few beings—at least on Earth— who can convert that tension into meaning.

You don't live in the universe. You live with it—by continuously resolving its ripple intent.

So before we close this chapter, remember: You are not small. You are not accidental. You are not separate from the system.

You are the point of convergence where ripple tension becomes light, sound, warmth, memory, emotion, action, choice, and consequence.

The next time you feel the warmth of the sun or hear your name spoken— know this:

That didn't "happen." It was resolved—by you.

1.8 Closing Reflection: Seeing Without Assumption

Now that you've walked with me through the illusions of light, sound, and energy— pause for a moment.

Look around you.

Everything you're perceiving—the color of the wall, the rustle of wind, the warmth of your breath— feels *real, present, happening.* And yes, it *is* real—but not in the way you've always assumed.

Reality is not what you're seeing. Reality is what is resolving—*around you, and because of you.*

Let that become your new compass.

You are not observing a world of objects in motion. You are witnessing a ripple script, one that's written in tensions and resolved in perception. Light isn't arriving. Sound isn't propagating. Energy isn't flowing. All of it is aligning, unraveling, and revealing—right at the edge of your awareness.

You don't need to imagine a magical alternative to modern science. You simply need to let go of the assumption that things are moving because they seem to change.

The illusion of motion has dominated human thought for centuries. But beneath it lies something more elegant: a still, structured universe, where ripple tension is the only reality— and all that you experience is the result of how that tension is resolved in your presence.

Embrace the Stillness

We have no need to chase particles. No need to visualize beams and currents and packets flying through voids. What we need is a deeper stillness. A stillness that allows us to see not *what is moving,* but what is aligning.

"The river does not flow. The banks shift. The sky does not shine. The field aligns. You are not watching reality—you are resolving it, ripple by ripple."

This stillness is not the absence of action. It is the source of all action—the canvas on which every illusion of movement is painted.

And this stillness is not far away. It is right here, within you. It is the ground from which you perceive, interpret, and participate in the universe.

What Comes Next

Now that the old ideas have cracked open, a new clarity begins to form.

In the next chapter, we will ask: Why does the world look the way it does—*to you*? Why blue sky, not infrared waves? Why music, not magnetic shifts? Why does *your* reality show up the way it does, in *your* field of perception?

We will discover that reality is not just *resolved around you*— it is also shaped by how you resolve it.

We will dive into the next ripple:

Why You See What You See.

Prepare not just to observe more deeply—but to understand the nature of seeing itself.

Chapter 2: Why You See What You See

Observer-Defined Perception | Reality as Ripple Alignment

You are not seeing the world—you are seeing what your field is able to resolve. There is no universal image, no fixed soundtrack, no shared canvas called "reality." What appears to you is not what exists—it's what aligns with the structure of your perception. Your body is not a passive receiver but a tuned ripple decoder, selecting and shaping what the universe reveals. In truth, you don't see what is—

You see what resolves.

And what resolves, becomes your reality.

2.1 Introduction – You Don't Just See, You Select

Imagine five people standing in the same room.

One sees a warm, cozy space filled with soft lighting and friendly energy. Another notices dust on the shelf and clutter in the corner. The third is lost in thought, barely aware of the surroundings. The fourth finds the atmosphere stifling and tense. The fifth feels inspired, moved by something they can't even name.

Same room. Same objects. Same ambient conditions. Five completely different realities.

What happened?

Did the room change five times? Did the furniture rearrange itself? Did the light flicker and moods shift?

No. What changed was what resolved—for each observer.

Perception Is Not Passive—It's Selective

We often speak as if perception is automatic. We open our eyes, and the world "enters." We hear a sound, and it "reaches" us. We feel something, and it "happens" to us.

But this is a mechanical myth. You are not a bucket collecting reality. You are a filter, a resonator, an interface—tuned to resolve certain ripple tensions in the universal field.

You don't just see what is there. You see what *you are aligned with.*

Just like a radio doesn't "hear" all frequencies, your body, your brain, your senses are tuned to a specific range— a narrow, precise band of ripple dynamics.

What falls outside this alignment remains invisible, inaudible, imperceptible. Not because it doesn't exist— but because you are not tuned to resolve it.

You Define What Resolves

Perception is not the act of receiving external reality—It is the consequence of internal alignment with ripple tension in your environment.

You are a participant, not a bystander.

Your eyes don't capture light; they resolve electromagnetic ripple patterns. Your ears don't absorb sound; they translate pressure ripple zones into meaning. Your skin, your intuition, even your attention—all are alignment mechanisms that shape what reality appears to be.

And what appears is not objective truth. It is selected resolution.

"You don't witness reality. You define the version of it that resolves in your field."

This is the shift that Chapter 2 will walk you through.

You will see how your senses are filters, not windows. You'll understand why no two observers ever see the same world, even when they stand side by side. And you'll come to realize that reality is not something out there to discover—but something within you, waiting to align.

Let's go deeper.

2.2 The Observer as a Ripple Filter

You are not just present in the universe. You are embedded within it—not as an object among objects, but as an instrument of ripple alignment.

Your body, your senses, your nervous system— these are not passive structures waiting to receive external input. They are finely tuned ripple filters, each built to decode specific forms of ripple tension present in the surrounding structured vacuum energy (SVE).

You are not "receiving information." You are resolving localized ripple stress into meaningful experience.

Your Body as a Ripple Interface

Every part of your body is designed to participate in ripple dynamics. From skin to retina, eardrum to spinal cord, you are a collection of field-aligned surfaces where ripple tension can collapse into perception.

This isn't symbolic. This is literal physics under UED.

You are not surrounded by the universe—you are woven into its ripple fabric, and wherever your structure aligns, you *see*, *hear*, *feel*, or *know*.

Let's break it down.

Eyes: Electromagnetic Ripple Resolution

What you call "seeing" is not the reception of photons. It is the alignment of your retina—a biological surface tuned to resolve ripple tension in a specific band of electromagnetic frequencies (roughly 400–700 nm). We call this "visible light," but it's not light in motion.

It's ripple stress collapsing across your visual field into a perception of color, shape, brightness, and form.

Everything outside that band—infrared, ultraviolet, X-rays—exists as ripple tension too, but you don't see them because you're not aligned to resolve them.

Ears: Pressure Ripple Resolution

Your auditory system isn't passively receiving sound waves. Your eardrum is a pressure-sensitive ripple membrane, tuned to detect tension shifts in the air between ~20 Hz to 20,000 Hz.

What we call "hearing" is simply the local resolution of ripple compression zones through your biological architecture. Your brain doesn't "receive" sound—it constructs it, after resolving ripple sequences into rhythm, pitch, and tone.

- No tension = no sound.
- Wrong alignment = no perception.

Your ear doesn't pick up noise—it translates field distortions into experience.

Skin: Thermal & Tactile Ripple Resolution

Your skin is a multi-modal ripple detection system. It resolves:

- Thermal tension (heat)
- Pressure tension (touch)
- Frictional variations (texture)
- Vibration (buzz, tremble)

When you touch something hot, you are not absorbing "thermal energy." You are simply aligning with a high-tension ripple zone and your skin's receptors resolve that stress into the sensation of heat.

The Nervous System: Interpreter, Not Messenger

The nervous system is often described as a signal highway. But under UED, it is better understood as a ripple logic interpreter.

Neural pathways don't merely "carry data"— they structure and decode ripple resolution patterns for integration in consciousness.

The brain doesn't sit inside the skull like a computer plugged into sensors. It is a live field-organism, processing ripple information as it collapses across bodily and cognitive alignment zones.

The nervous system isn't relaying information. It is translating ripple tension into experience in real time.

You Are the Filter, Not the Window

What you see, hear, and feel are not properties of the universe. They are properties of your ripple configuration.

Reality is not something you look through—it's something you build. You don't "observe" the world. You resolve a version of it, shaped by the tuning of your field.

You are not a passive observer. You are the ripple-filtering lens through which existence becomes meaningful.

2.3 The Tuning Principle: Why You See One Version of Reality

If the universe is filled with ripple tension— a silent, structured stillness waiting to be resolved— then why do you see only one version of it?

Why not the full field? Why can't you hear beyond 20 kHz? Why don't you see X-rays or feel magnetic flux directly? If the field is universal, shouldn't perception be infinite?

The answer lies in the principle of tuning.

You don't experience reality in its totality. You experience a specific band of ripple resolution, because of how you—your body, brain, senses, and mind—are tuned.

1. Your Structure Determines What You Resolve

At the most fundamental level, your perception is shaped by how you're built.

- Your retina responds to electromagnetic ripples between ~400–700 nm
- Your eardrum detects air pressure ripple fluctuations between ~20–20,000 Hz
- Your skin resolves ripple tension across pressure, thermal, and vibrational thresholds
- Your olfactory and gustatory receptors respond to specific chemical ripple patterns (smell and taste)

Everything beyond these bands—gamma rays, infrasound, ultraviolet, magnetic waves—exists in the same field, but you are not structurally aligned to resolve them.

That doesn't mean they're not real. It just means they're real in a way that's invisible to you.

Perception is not limited by reality. It is limited by your ability to resolve it.

2. Biological Tuning: Nature's Ripple Filters

Evolution didn't gift you the full range of perception. It gave you what was useful for survival—and nothing more.

- Bees see ultraviolet ripple patterns on flowers.
- Bats resolve ripple pressure patterns far above human hearing.
- Snakes detect infrared ripple tension as "heat vision."
- Birds navigate using magnetic ripple alignments in the Earth's field.

Each species resolves a different ripple layer of the same universe.

There is no single "reality." There are many overlapping fields, and each living system sees only what it's built to decode.

3. Cognitive Filtering: The Mind's Ripple Gatekeeper

Even within your biological limits, you don't perceive everything. You perceive what your attention selects, what your memory compares, and what your bias allows.

- You tune out background noise to focus on a voice.
- You miss the details your brain doesn't label as important.
- You reinterpret ripple data based on emotional state, belief, expectation.

This mental layer doesn't just filter information—it reshapes how ripple resolutions are experienced.

Two people can hear the same sentence and come away with two different truths. Not because the sound was different—but because their cognitive filters resolved the same ripple pattern differently.

You don't just see what exists. You see what your filters allow to form.

Perception = Resolution Bandwidth

Think of yourself as a tuned instrument in a giant ripple orchestra. You're not hearing the whole song—you're catching the parts you resonate with.

Color is not in the world—it's in your field of resolution. Sound is not "out there"—it's the way your brain resolves ripple compression. Texture is not a surface property—it's a pressure differential resolved across your skin. Time is not a cosmic flow—it's a mental rhythm of changing alignment states.

The world does not have color. You do. The world does not speak. You hear it. The world is not in motion. You are resolving change.

You Are a Localized Ripple Decoder

There is no such thing as "neutral perception." Every experience is filtered through:

- Your physical build

- Your sensory bandwidth
- Your mental patterns
- Your alignment with ripple fields at any given moment

And therefore:

What you call reality is not a mirror of the universe— It is the version of the ripple field your tuning allows to collapse into meaning.

This is not limitation. This is precision.

Now that we understand the nature of perceptual tuning, the next question is inevitable:

If every observer is resolving a different version of reality, is there any common ground?

Can we ever claim objectivity in a world that is observer-aligned?

To explore that, we now turn to the tension between objective ripple fields and subjective resolution—and what it really means to live in a shared world.

Let's move forward.

2.4 Ripple Alignment vs. Objective Reality

For centuries, science has pursued an idea that sounds noble, clean, and absolute: the objective world.

An external universe, separate from the observer. Full of measurable quantities, repeatable events, and consistent laws—existing whether we look at it or not.

It gave rise to objectivity, empiricism, rationality, and modern physics. And to be fair—it served us well.

We could build machines. Predict outcomes. Launch rockets. Cure diseases. All by assuming that the universe is out there, and we're simply measuring what's already happening.

But now, under Universal Energy Dynamics (UED), we step beyond that assumption— not to abandon science, but to complete it.

Because the truth is this:

Yes, objective ripple fields exist. But they are meaningless—until resolved by an aligned observer.

The Field Is Real—But Incomplete

The structured vacuum (SVE) is full of ripple tension. There are gradients, loops, compressions, distortions, and alignments. They exist with or without you. But until something aligns with those patterns— until a decoder is present—they don't become *color*, *sound*, *heat*, *form*, or *meaning*.

They are raw potential, not reality.

Just as a musical score is not a song until someone plays it, the ripple field is not "reality" until someone resolves it.

You Are the Other Half of the Equation

The ripple field is the script. You are the interpreter. Reality only "happens" when those two meet.

This is not a philosophical fantasy. This is the operational truth of perception.

A beam of sunlight is just a ripple tension zone. Your retina resolves it into brightness, direction, shape. Remove the observer—remove the resolution— and there is no light. No day. No scene.

The ripple remains. But the world does not appear.

- No observer → no resolution
- No resolution → no experience
- No experience → no perceived "reality"

Classical Science: External Reality

Traditional physics assumes there's a fixed universe "out there," waiting to be discovered like a book already written.

That's why we chase objectivity— repeatability, measurement, detachment.

But this method forgets the most important component: Every measurement is a resolution. Every observation is an interaction between ripple field and the tuning of the observer or instrument.

There is no perfectly objective view. Only structured field phenomena filtered through local alignment.

Reality Is Co-Authored

UED doesn't reject objectivity. It reframes it.

Yes—the ripple field is real, patterned, and structured. But it is not reality until an interface with the right tuning collapses those patterns into something meaningful.

Reality is not "out there." It is not "in here." It is between—born from the interaction of ripple tension and alignment structure.

You don't observe reality. You co-create it—by resolving tension into perception.

This Is Why Truth Is Layered

If perception depends on tuning, and if tuning varies across observers, then "truth" becomes a layered construct— different observers resolving different bands of the same field.

It explains:

- Why bees see ultraviolet patterns we don't
- Why machines detect radiation we can't
- Why spiritual experiences differ across cultures
- Why two people interpret the same event differently
- Why no single "objective truth" has ever satisfied the entire world

There is one field, but many observers. And thus, many resolved realities—each valid within its alignment window.

So next time someone asks, "But what's really out there?" you can smile and say:

"Ripple tension. Waiting for someone to show up—and turn it into reality."

2.5 Why Different Observers See Differently

Reality is not one-size-fits-all.

Stand in a garden at noon.

You see green leaves, a yellow sun, and the shimmer of heat rising from stone. A bee beside you sees ultraviolet patterns on petals that are invisible to you. A bat flying nearby hears none of this—but detects high-frequency ripple tension you can't even imagine. A thermal camera mounted on a drone registers none of the above. It sees gradients of infrared heat, mapping the ripple field in wavelengths beyond your biology.

Five versions of the same space. All different. All valid.

What changed?

Not the field. The field is shared. What changed is who's resolving it—and how.

Different Tunings, Different Realities

Every organism—every machine, for that matter— is tuned to a specific bandwidth of ripple alignment.

- A bee resolves electromagnetic tension in the ultraviolet band (~300–400 nm). It sees flower patterns you will never see with your naked eye.
- A bat resolves air-pressure ripple tension at ultra-high frequencies (up to 120 kHz). It maps space through echolocation, not light.
- A snake detects thermal ripple changes through pit organs. It "sees" heat—infrared resolution through specialized receptors.
- A human sees light from 400–700 nm, hears sound from 20–20,000 Hz, and resolves touch, smell, and taste in

corresponding zones.

Same ripple field. Different decoding instruments. Different experienced realities.

Reality Is Not Imagined—It's Resolution-Specific

We often say "perception is subjective," as if it makes perception less trustworthy. But under UED, subjective does not mean imaginary.

It means: The version of reality you experience is the version your ripple configuration allows to resolve.

Just because a bat can't see what you see doesn't make its world unreal. Just because you can't hear what a dolphin hears doesn't make your reality incomplete.

There is no master reality, just as there is no master key to every lock. There is only the shared ripple field—and each observer aligns with a different part of it.

Machines as Observers

This isn't limited to biology.

Machines "see" infrared, ultraviolet, X-rays, magnetic fields. Not because they're conscious—but because their sensors are tuned differently.

Your phone's night camera can pick up details your eyes miss. A radio telescope detects ripple patterns stretched across light-years of space— waves that no human organ could ever register directly.

But even machines aren't passive. They too require interpretation, translation, and decoding, often by another observer—you.

So while their tuning differs, the principle remains the same: All perception is a function of alignment.

Implication: Truth Is Layered, Not Singular

Truth is not a single output from a cosmic machine. It is a ripple event, experienced through individualized resolution systems.

That's why:

- Science evolves: new tools = new perceptions
- Religion varies: different minds = different alignments
- Art flourishes: unique inner configurations = unique outer interpretations

There is one universal ripple field. But infinite ways to meet it.

So when someone says, "Reality is subjective," you can now say:

"No—it's alignment-specific." And alignment is no less real than structure. It is the *means* by which structure becomes meaningful.

In the next section, we'll explore how conscious attention further shapes ripple resolution— and how tuning can be refined, focused, and even expanded by the willful observer.

Let's move deeper into perception as dynamic engagement.

2.6 Conscious Attention = Field Precision

Perception doesn't end with the senses. In fact, your senses are just the beginning.

While your body may be tuned to certain ripple bands biologically— your attention decides what resolves with precision.

You see with your eyes, yes. But you perceive with your focus.

Where you place attention, you place alignment. And where you align, resolution becomes sharper, richer, deeper.

Attention as Ripple Alignment

Have you ever noticed how something you weren't aware of suddenly becomes clear when you focus?

- The ticking of a clock you didn't hear until the room went quiet
- A faint scent you detect only after pausing and breathing deeply

- An idea that was buried in background thought, which surfaces the moment you concentrate

This is not psychological—it's field dynamics.

Your attention functions as an alignment amplifier. It narrows the ripple filter and sharpens your tuning to a more specific range.

In UED terms:

Conscious attention increases the coherence of your internal ripple receptors, allowing finer gradients of the field to resolve into meaningful experience.

Meditation: The Art of Still Alignment

In meditation, you are not "doing nothing." You are retuning your internal ripple field.

By quieting mental noise—distracting ripple chatter—you allow subtle field dynamics to emerge. Thoughts become clearer. Awareness broadens. Often, things that were previously invisible—patterns, tensions, inner truths—become suddenly apparent.

Why?

Because you are no longer flooded with scattered ripple interactions. You are holding a stable alignment, which allows lower-frequency, deeper ripple layers to resolve.

Imagination and Creativity: Inner Ripple Engineering

When you imagine, you don't disconnect from reality. You realign your inner field with non-local ripple possibilities.

Creativity is not about invention. It's about accessing uncollapsed configurations of the ripple field and giving them new form.

Artists, inventors, and visionaries don't create from scratch. They resolve invisible ripple tensions into visible form—in music, design, writing, mathematics.

Imagination is ripple play. Creativity is ripple resolution beyond sensory limits.

Insight and Intuition: Sudden Alignments with Deeper Fields

Sometimes you "just know." You don't analyze. You don't reason. It just hits you.

This is intuition—not magic, not guesswork, but a moment when your inner field suddenly aligns with a deeper ripple structure in your environment, your memory, or your subconscious.

It is a non-linear collapse of ripple tension into immediate understanding.

In that moment, the precision of resolution bypasses conscious steps. What takes hours through logic resolves in a flash through deep field resonance.

Awareness Is the Real Instrument

Your biology is your hardware. But your awareness is the interface that controls precision. What you are aware of, you sharpen. What you ignore, you blur.

And this has profound implications:

- Two people with the same eyes may see different truths—because one is paying attention.
- Two thinkers with the same knowledge may draw different conclusions—because one is aligned.
- Two artists with the same tools may create vastly different work—because one is tuned deeper into the field.

Attention is alignment. Awareness is precision. And consciousness is the field's way of fine-tuning itself.

So the next time you pause, breathe, and direct your awareness to a feeling, a sound, a thought—know this: You are not merely observing.

You are realigning the ripple field within you, allowing reality to collapse in higher resolution.

And in that collapse, clarity is born.

2.7 There Is No "Outside World" Without a Decoder

We're trained to think that the world exists "out there"— as a ready-made container of objects, sounds, colors, and events—waiting for us to open our eyes and discover what's already happening.

But what if that entire picture is backwards?

What if the so-called *outside world* only becomes *world* when an observer—a decoder—is present to resolve it?

Let's revisit the most radical (yet obvious) realization from UED:

Nothing exists experientially until it is observed. No perception means no reality—not in any meaningful sense.

Structure Exists—But Experience Requires Resolution

Yes, the ripple field is always there. Yes, tension gradients exist whether or not you're watching. But light is not light without a decoding interface. Sound is not sound without a pressure-sensitive receiver. The sky is not blue without a biological visual filter. Heat is not hot unless ripple tension resolves on your skin.

The universe is full of unresolved ripple potential—but until something collapses that tension into form, there is no *world as we know it.*

Sound in the Vacuum: A Classic Example

A rocket explodes in space. On Earth, we would expect a thunderous boom. In space, we hear nothing.

Why?

Not because "sound disappears." But because there is no medium to align ripple pressure. No air. No compression. No eardrum. No decoder.

The explosion still creates ripple tension—but with no aligned interface to resolve it, nothing happens. No experience. No sound. No moment.

This is true for everything.

The Sky Isn't Blue—You Are

The sky doesn't hold a color. It holds a field of unresolved electromagnetic ripple patterns, which your retina and visual cortex resolve into blue.

Change the tuning—like in bees or infrared cameras—and the sky becomes something else entirely.

So where is the "real" sky?

It's not "out there." It's in the interaction between ripple field and observer.

There is no color in the world. There is only alignment in your decoding system.

Pain, Pleasure, Beauty—None Exist Without You

A knife touches steel—nothing happens. The same knife touches skin—ripple tension resolves as pain.

Sunlight falls on rock—it warms. Sunlight falls on skin—it burns, or comforts. Sunlight falls on chlorophyll—it nourishes.

Same field. Different interface. Different outcome.

Meaning does not belong to the universe. Meaning arises only within the observer, as ripple resolution collapses into emotion, memory, sensation, or understanding.

You Are the Converter

You are not a lens. You are the converter—turning tension into texture, structure into sensation, possibility into perception.

Without you, there is no music, no beauty, no sadness, no color. There is only the silent hum of uncollapsed ripple potential.

That's why UED doesn't say "reality is illusion." It says:

Reality is unresolved structure—until someone aligns and decodes it.

Meaning Emerges in Your Field

The universe is not trying to "tell you" anything. It is not broadcasting messages, sending packets, or pushing signals. It is simply holding structured tension across infinite scales.

The meaning, the message, the moment— arises when your internal field collapses that tension into form.

Reality is not given. It is evoked—ripple by ripple, alignment by alignment, within the perceiver's field.

As we prepare to close this chapter, we now stand face-to-face with the ultimate realization:

You are not discovering the universe. You are constructing the version of it that resolves through you.

In the final section, we'll embrace this fully:

You are not just seeing reality. You are participating in its formation.

Let's end this chapter where the observer becomes the artist of perception.

2.8 Closing Reflection – Perception Is Participation

You've been told all your life that you're looking at the world. That your senses are windows. That your mind is a mirror. That reality enters you from the outside.

But now you see the deeper truth:

You are not looking at the world. You are activating it—ripple by ripple—through alignment.

Every moment you exist as an observer, you are turning ripple tension into form, into feeling, into fact. You are not watching a movie—you're projecting it. You are not reading a script—you're collapsing it into speech.

Perception Is Not Passive—It's Participatory

To perceive is not to receive. To perceive is to participate.

Your biology selects the band. Your attention sharpens the resolution. Your memory filters what matters. Your emotions color the outcome. And your awareness determines which part of the ripple field becomes real—for you.

That is not observation. That is co-creation.

The World You Know Is the One You Resolve

There is no *default* reality waiting for you to observe it. There is only a structured field, silent and unread, until you tune in.

Once you do, what appears as shape, sound, color, movement, meaning— is not what *is*. It is what aligns.

You don't see what is. You see what aligns. And what aligns, becomes your truth.

Two people can live in the same house and inhabit different universes. Two scientists can look at the same data and tell different stories. Two minds can read the same sentence and find different meanings.

Because perception is never neutral. It is ripple selection.

And selection is not observation—it is a creative act.

You Are a Living Decoder

You are not a spectator in the ripple field. You are an interpreter. A sculptor. A mirror that bends the unseen into form.

Every breath, every thought, every sensation is a moment of participation in the ripple field.

Even silence is not empty. It is alignment with the unresolved.

Even confusion is not a void. It is misalignment waiting for clarity.

And truth is not static. It is what your field is ready to resolve—at this moment, in this state.

The Truth Is Aligned, Not Absolute

So let go of the idea that truth is something "out there" to be captured like prey. Truth is not a fixed object in space. It is the outcome of resolution—within your field, with your tuning.

When the alignment is sharp, you see clearly. When the field is noisy, the truth is distorted. When the ripple settles, reality speaks. And when you change, the world you see changes with you.

Closing Invitation

As we close this chapter, one idea should stay with you—not as a concept, but as a way of being:

You are not seeing the world. You are shaping the visible from the invisible. You are participating in reality by the way you align.

In Chapter 1, you learned that light, sound, and energy are not entities—but consequences of ripple resolution. In Chapter 2, you now understand why you see what you see—not because it's there, but because you are ready to resolve it.

Now we must ask: What is this present moment, this ripple configuration we call *now*? Is time really flowing? Or are we aligning with change in a deeper, more still dimension?

Let's explore that next.

Then, Who Is Seeing?

— You Are That

At the end of every question—Who is seeing? Who is perceiving? Who is resolving this reality?—one answer remains:

You.

But not the personal "you." Not the name. Not the history. Not the identity you picked up over time. Not the body that changes, nor the mind that drifts.

No. This You is deeper. It is not a person watching the world. It is the point of alignment where ripple tension collapses into clarity.

You are the place where the formless becomes form. You are the silence that allows meaning to echo. You are the ripple interface through which the universe speaks itself into experience.

Not outside. Not inside. Just *That.*

You Are Not Looking at the World — You Are Its Decoder

The world you see is not arriving—it is resolving. And it resolves because *you* are aligned.

You are not a bystander. You are the resolution event.

The field waits silently— until You appear, and it unfolds into light, sound, feeling, and thought.

You are not seeing. You are *being* the seer.

Know Thyself — As the Resolver

The sages said "Know Thyself." The physicists searched for a Theory of Everything. The philosophers sought the observer behind all observation.

But under UED, all paths meet at one insight:

"The universe becomes what you are ready to resolve." And the one who resolves it is That— not a self, not an ego, but the conscious alignment through which structure becomes sensation.

This is not mysticism. This is not abstraction. This is the ultimate field alignment truth:

You are not part of the universe. You are the place where the universe becomes part of itself.

Chapter 3: The Present Moment Is the Only Truth

Memory, Imagination, and the Ripple of Now

You think time is passing. That the past is behind you, the future ahead, and the present just a fleeting moment in between. But what if that's just a story told by memory and anticipation? In truth, nothing exists except what is resolving now. There is no river of time—only a still field of ripple tension, where the present is the only point of active collapse. Everything else—past or future—is either an echo or a possibility, shaped by your alignment in this exact moment.

3.1 Time Isn't What You Think

You've been told time moves forward. That it flows like a river—steady, irreversible, always pulling you into the future.

You've measured it with clocks. Celebrated it with calendars. Feared its speed and chased its scarcity.

But what if that's all a deeply convincing illusion? What if time doesn't move at all?

Let's break the spell.

The Grand Illusion: Time as a Flow

In the classical view, time is seen as a dimension of motion—a linear thread connecting events from past to future. An object moves → change happens → time is said to pass. Your memories pile behind you; your plans stretch out ahead. The clock ticks, the sun rises and sets, and you move through life thinking time is carrying you with it.

But under Universal Energy Dynamics (UED), this idea collapses.

Time is not flowing. You are not moving through time. Instead, ripple fields are shifting—and you're resolving them, moment by moment.

There is no stream of time. There is only a still field of ripple tension—where some parts are in active alignment with your system, and some are not. What you call "now" is the only place where that alignment becomes experience.

No Past, No Future—Only Resolution

The past does not exist. It is a pattern of resolved ripple configurations, stored as alignment echoes in your memory field. When you remember, you don't "go back in time." You simply replay a previous resolution pattern in your present field.

The future does not exist either. It is unresolved ripple potential—a field of tension that has not yet aligned. When you imagine, plan, or worry, you are projecting your current structure onto unresolved ripple zones and calling it "what may come."

The only thing that truly exists is what is resolving right now. Not what was. Not what might be. Only the ripple tension that is actively collapsing in your field of perception—this moment, this breath, this choice.

Time = Alignment, Not Motion

Let go of the idea that time is a conveyor belt carrying you forward.

Instead, imagine a field—silent, complex, unchanging. You are not walking across it. You are tuning into ripple pockets, and each alignment feels like a "moment." As your tuning shifts, different ripple regions resolve. And the sequence of those resolutions becomes your experience of time.

You are not flowing through time. You are sequencing ripple collapses, and that sequencing *feels like* time.

Why It Feels Like Time Moves

Because your mind tracks change. Because your brain layers moments into memory. Because your attention shifts with rhythm and motion.

But these are not markers of time passing. They are internal responses to ripple realignment.

The sun doesn't rise—it realigns. Your body doesn't age in time—it resolves ripple tension through biological processes. The clock doesn't track time—it measures regular intervals of physical field change, which you've agreed to call "seconds" and "minutes."

But none of this means time is moving. It only means alignment is happening, and you are present enough to perceive it.

Conclusion: Time Is Not a Thing You Travel Through

It is not a container. Not a force. Not a stream.

Time is a feeling of difference between successive ripple resolutions. The rest—past and future—is imagination and memory, both dependent on your alignment right now.

So if you're searching for truth, you won't find it in the past. And you won't create it in the future. You'll only encounter it where ripple tension collapses into meaning—here, in the present.

Time isn't what you think. It's not a road. It's a ripple that resolves—only when you're tuned.

3.2 Time as Ripple Perception, Not Linear Flow

Time has been described as many things—a dimension, a flow, a stream, a force, even a ticking cosmic clock. But what if it's none of these? What if time is not *out there*, but something that arises *in you*, through your interaction with ripple fields?

Let's drop the metaphor of movement. Let's stop imagining time as a river you're drifting through. Instead, see yourself standing still—while ripple tensions shift, align, and collapse all around you. And what you call "time" is simply the sequence in which you perceive those collapses.

Not a Force, Not a Flow

Time doesn't push you. It doesn't carry events from past to future. It doesn't flow like water or radiate like heat.

In the ripple-based reality of UED, nothing flows at all. There is no substance called time. There is only a structured vacuum filled with ripple tensions—zones of alignment waiting to collapse into experience. And what you experience as "passing time" is merely your tuning shifting across this silent field.

"Time isn't a thing. It's a coordination of ripple collapse—filtered by you."

Time as a Sequencing Experience

Imagine a series of dominos set up across a floor. You don't see them all fall at once. You experience them *one after the other*, in a sequence. Not because they exist in time—but because the sequence itself *creates* your perception of time.

Likewise, ripple resolutions happen continuously across the universe, but you only perceive them in a certain order, filtered through your body, your senses, your awareness.

Time = The inner rhythm of how your field resolves external tensions.

Each moment is a unique resolution pattern—a collapsed segment of universal potential aligned with your current state. String enough of those resolutions together, and your mind calls it "time passing."

Shifting Configurations, Not Flowing Moments

You're not "traveling" from one moment to the next. Instead, your field is dynamically tuning, and different ripple configurations come into resonance with your structure.

It's not the moment that moves. It's *you*—as a field of perception—who shifts into alignment with the next ripple pattern.

And since each alignment produces a slightly different experience, the illusion of flowing time is born.

But in truth...

"You're not moving through time. You're resolving through change."

Tuning Bandwidth = Time Bandwidth

Your sense of time is limited by the resolution range of your biology and awareness. You can't see microseconds without special instruments. You can't feel geologic epochs. Your "timeline" is simply the range of ripple transitions your nervous system and consciousness can handle.

Different species live on different timelines—not because time varies, but because their tuning bandwidth is different.

Even in your own life: when you're fully immersed, time seems to vanish. When you're anxious or distracted, time drags. That's not time changing. That's your alignment shifting.

The Big Shift: Time as Internal Mapping

So instead of treating time as something "out there" measuring you, realize: It's your mapping of ripple change—an internal construct built from resolution order.

There is no external "clock" governing the cosmos. There is only ripple tension—and the observer's pattern recognition of its collapse.

What you experience as duration, memory, anticipation—it's not because time is moving. It's because your alignment is dancing. And your awareness is *writing the choreography*.

Summary Thought

You don't "have time." You don't "spend time." You don't "lose time."

You experience shifting ripple realities, one alignment at a time. And that unfolding rhythm? That's what you call your life.

"Time is not your path. It's your pattern of resolution."

3.3 Memory: Ripple Echoes of Prior Alignment

Memory is often misunderstood as a rigid archive—a mental vault storing snapshots of the past, like scenes in a film reel waiting to be played back. But under the ripple framework of UED, memory is not a fixed recording. It is a stored resolution pattern—a residual echo of how your field once aligned with experience.

When you recall a memory, you're not pressing 'play' on a file. You are re-resolving a former pattern within your current field configuration. This means every memory is slightly different each time it resurfaces—because *you* are different. The recall process is not retrieval—it is reinterpretation.

Memory Is a Ripple Echo, Not a Replica

The brain doesn't store reality—it stores tension blueprints. The mind doesn't retrieve the truth—it reconstructs coherence. What we call memory is the field's attempt to recreate a former harmony using the current state of the ripple fabric. That's why even deeply factual memories—like dates, names, or locations—can feel different depending on your emotional or energetic state.

Emotional Memory = Resonant Tension

Emotionally charged memories leave stronger ripple impressions. Why? Because they involved greater energetic displacement—more significant ripple realignments. Trauma, joy, grief, love—these aren't "more important" memories; they're denser ripple events, producing deeper field indentation.

This is why:

- A scent can bring tears.
- A song can resurrect heartbreak.
- A gesture can reawaken long-forgotten warmth.

These aren't just chemical associations—they're vibrational resonances, aligning your current field to a prior tension profile.

Memory Is Not the Past—It's the Present Interpreting the Past

Every act of remembering is an act of present alignment. That's why:

- The same childhood event can feel different at age 10, 30, or 60.
- Healing happens not when we erase memories, but when we realign how we hold them.
- Memory is fluid—malleable to awareness, not bound to accuracy.

Memory becomes wisdom not when it's preserved, but when it's recalibrated through conscious interpretation.

"You don't carry your past. You carry your unresolved attempts to understand it."

—Chavan's Ripple Principle of Memory

3.4 Imagination: Future as Non-Collapsed Resolution

You close your eyes and see a version of tomorrow.

Maybe it's a meeting. A sunrise. A possibility. Maybe it's an invention, a conversation, a dream you've never told anyone.

But here's the truth: that future does not exist. It is not waiting for you. It is not hiding in a timeline ahead. It is not a location. It is not real—yet.

The future is unresolved ripple tension—a field of possibility that hasn't yet aligned with your structure.

And imagination? That's your tuning probe, scanning the ripple fog for what *could* be.

Imagination as Alignment Projection

What happens when you imagine?

You don't pull images from thin air. You project alignment patterns forward from your current state into a field of unknown resolution. You test what may fit. You rearrange tension. You sculpt options.

Imagination is not fantasy. It is field experimentation.

"To imagine is to pre-align—mentally—to that which hasn't yet collapsed."

The child imagining themselves flying is reconfiguring inner ripples to explore the feel of weightlessness. The scientist picturing a new theory is aligning with a field of unknown structure and mapping ripple resolution through thought.

Planning = Structured Pre-Collapse

Planning isn't a timeline. It's an ordered ripple simulation—a sequential alignment of imagined outcomes that feels like a strategy.

When you plan a trip, build a product, or chart a decision, you're not moving into the future. You're creating a ripple corridor that increases the chances of favorable resolution.

Planning doesn't guarantee success. But it preconditions your field—it *prepares alignment zones* that are more likely to collapse the way you want.

This is not mysticism. This is practical ripple engineering.

Creativity: Accessing Unused Ripple Tension

Where does art come from? Music? Invention? New ideas?

Not from the "future," but from untapped ripple configurations around you—patterns you haven't resolved yet.

The more flexible your internal field, the more permutations of ripple tension you can access. That's creativity.

Creativity is not creating from nothing. It's resolving tension in unexpected ways. It's feeling the edge of what's unresolved and daring to shape it.

There Are No Future Events

When you say, "It will happen tomorrow," what you really mean is: "If ripple configurations align in a certain way based on my current structure, then I will experience a resolution that feels like that event."

There is no "event" ahead. There is only an open ripple field—fluid, uncommitted, and unresolved.

"The future is not ahead. It is around. And your tuning decides what becomes real."

Collapse Happens Now

The only time anything ever happens is now. You do not "reach the future." You collapse a potential into the present.

That's all a "moment" is: A ripple configuration snapping into alignment with your observer field.

All imagination does is stretch your reach—like a feeler, touching ripple tensions that haven't yet come into resonance.

Summary Thought

So the next time you say, "I can't wait for tomorrow," remember: There is no tomorrow waiting.

There is only you—now—deciding what to align with next.

"Imagination is not escape. It is ripple rehearsal. The future is not in front of you. It is *becoming*, through you."

3.5 The Now – The Only Place Ripple Resolution Happens

Close your eyes. Breathe. Feel your heartbeat. Notice the warmth around you, the sounds, the thoughts passing by. This moment—this very now—is all that's ever actually happening.

Everything else...the yesterday you recall, the tomorrow you anticipate— are either echoes or projections.

Only the present is real in the truest sense, because only now do ripple tensions resolve into experience.

Resolution Happens Only in the Present

In the framework of Universal Energy Dynamics (UED), reality is not made of flowing matter or traveling forces— it is made of structured vacuum energy hosting ripple tensions—zones of misalignment trying to resolve.

And resolution? It doesn't wait. It doesn't queue up in a timeline. It doesn't happen in "past tense" or "future tense." It can only occur in the now—the only field where your observer structure, your attention, and the ripple configuration meet.

"The now is not a moment in time. It is a convergence point of field and awareness."

Past: Echoes of Resolution

What is memory?

It is not you going back in time. It is your field revisiting the echo of a past resolution pattern—retrieving tension traces that *once aligned.*

The past is not gone—it's just no longer resolving. What remains is a field residue, a signature in your nervous system, a recording in your field alignment history.

You cannot change the past. But you can realign your interpretation of it. Because even memory—like all perception—resolves only in the now.

Future: Unresolved Alignment

The future doesn't exist yet.

What you call "the future" is nothing more than ripple tensions still in the field, waiting for the right alignment. You may sense them. You may imagine them. You may even predict them.

But they haven't resolved. They're still potential. They're not real until now collapses them into experience.

"The future is not ahead. It's ambient—floating in ripple tension until resolution brings it home."

The Present: The Only Active Zone

Let's be clear:

- No perception happens in the past.
- No experience happens in the future.
- Everything you know, do, feel, create, change—happens only now.

That's not philosophy. That's ripple mechanics.

The present is not a timestamp. It is the active zone of resolution. It is where ripple tensions meet a receptive observer and collapse into experience.

Without your presence, tension remains unresolved. Without ripple tension, your presence has nothing to decode. Together—they create *now*.

Living in Resolution

When you feel aligned, centered, "in the zone"— what's really happening?

Your observer field is dynamically synchronized with ripple tensions resolving in real-time. You're not just experiencing the now—you're *inhabiting it fully.*

That's flow. That's mindfulness. That's being alive.

"You're not moving through time. You're standing in a still field, where the universe resolves itself through you."

Summary Thought

You cannot change what has resolved. You cannot live in what hasn't. You can only interact with what's resolving right now.

"You can only experience what is resolving. And resolution only happens now."

3.6 Why Mind Escapes the Present

You are here. But the mind—it's already gone.

Gone to a memory, a plan, a fear, a fantasy. Gone to a version of you that existed before, or one you're trying to become. Gone anywhere but now.

Why does the mind escape the present? Because now is unfiltered. Raw. Exposing. And the ego doesn't like that.

Ego Seeks Time, Not Truth

The ego—the constructed identity of "I"—is not grounded in presence. It's built from accumulated ripple patterns: Stories of the past. Ambitions for the future. Comparisons, attachments, validations.

It doesn't thrive in real-time alignment. It thrives in narratives—where it can shape, defend, and preserve itself.

So it seeks time as a buffer zone:

- Regret gives it roots.
- Expectation gives it direction.
- Distraction gives it escape routes.

Time becomes the ego's stage. Now is too small, too uncertain, too vulnerable.

"The ego needs time to exist. The soul needs only now."

Ripple Misalignment: The Root of Psychological Struggle

When you're not fully in the present, you're not fully aligned. Your field starts interacting with ripple echoes or projected tensions—not real-time resolution.

This causes:

- **Anxiety** – from unresolved future ripple scenarios
- **Nostalgia** – from longing for past alignments
- **Restlessness** – from inner ripple noise seeking grounding
- **Overthinking** – from trying to resolve what isn't resolving yet

You're not broken. You're just misaligned.

The system isn't malfunctioning—it's simply not centered.

Attentional Grounding: Staying in the Resolution Zone

So how do you return?

By grounding attention—the conscious anchor of your field—into the now.

Not by force, but by *participation*:

- Feel the warmth of your skin.
- Hear the subtle ambient sounds.
- Observe thoughts without following them.
- Breathe and sense the breath—not as an idea, but as a ripple resolution event.

"Attention is not just focus. It is your tuning knob to the field."

The more you attend to real-time ripple resolution, the more you *become present*. And the more present you are, the less power the mind has to escape.

Presence Is a Practice of Perception

Being present isn't stillness. It's *dynamic tuning*—like a surfer on an active wave of reality. It's continuous, fluid participation in resolution.

You can't stop thoughts. But you can recognize that they're just ripple echoes or projections. And you can choose not to align with them.

Presence is not absence of mind. It is right-mindedness—aligned, aware, alert.

Summary Thought

You escape the present not because you're flawed, but because the constructed self fears disappearing in the rawness of now.

But you are not the mind. You are the one perceiving the field. And the field is always, only, resolving now.

"The mind escapes to survive. Awareness returns to align. You are not your escape—you are the field waiting to be felt."

3.7 Stillness Is Now: The Gateway to Alignment

When you hear the word *stillness*, what comes to mind? A quiet forest? An unmoving lake? An absence of motion?

But stillness, in the field of Universal Energy Dynamics, isn't a lack of movement— it's the zero-point of ripple conflict. It's not inactivity—it's pure alignment.

Stillness is not about *doing nothing.* It's about *being everything that is resolving now*—with no resistance.

Stillness = Full Presence in Resolution

You are always in the ripple field. But your attention drifts—pulled by echo tensions of the past or projected waveforms of the future.

Stillness is when all ripple receptors in you align with what's happening *now*. Not yesterday's regret. Not tomorrow's anticipation. Just the present unfolding in its raw, unfiltered precision.

That's why stillness feels sacred. Because it's not the absence of sound, thought, or sensation— It is being fully tuned to what *is actually resolving* in this moment.

Meditation, Surrender, Awareness

What do all contemplative practices—from meditation to prayer to deep surrender—have in common?

They don't *add* anything. They help you withdraw from misalignment.

- Meditation teaches you to observe without clinging
- Surrender teaches you to stop resisting ripple flow
- Awareness teaches you to expand your tuning bandwidth

Each one is a technique to let resolution happen, instead of distorting it through ego, memory, or mental chatter.

"Stillness is not a practice. It is a return. A return to the natural state of ripple harmony within the now."

Why Clarity Emerges in Stillness

Ever notice how your best ideas arrive when you're not trying?

In the shower. On a walk. In a moment of pause.

That's not a coincidence. That's ripple field alignment reestablishing itself.

In stillness:

- Noise reduces
- Field interference drops
- Ripple resolution becomes sharper

And suddenly, you *know*. Not because you figured it out, but because the field resolved clearly enough for you to perceive.

Truth, clarity, insight, intuition—they are not products of thought. They are products of precise stillness.

Stillness Is Dynamic, Not Dead

Let's bust a myth: Stillness doesn't mean freezing or emptying the mind into a void.

It means being fluidly aware—moment by moment—of how resolution is happening.

It is:

- Alive
- Sensitive
- Present
- Quietly intelligent

It's the active silence where all things stabilize.

"Stillness is not silence. It is the hum of perfect alignment."

Final Thought: The Gateway Is Always Open

Stillness is not a destination. It's a return point—always here, always accessible.

You don't need to change the world to find peace. You just need to return to what is resolving—now.

That is where truth lives. Not in beliefs. Not in theories. But in present resolution, inside a still, receptive observer.

"Stillness is the key. The field is the lock. Alignment is the opening."

3.8 Closing Reflection – Be Where the Resolution Is

Stop running. There is no clock chasing you. No timeline you need to outrun. No cosmic deadline you're failing to meet.

Time is not a river you're being swept away in. It is a field resolving in real-time—and you are its axis of perception.

You are not early. You are not late. You are not off-course. You are exactly where resolution happens—because the field only resolves through presence, not projection.

You Don't Have to Chase Time

The illusion of time seduces you into hurrying. You feel the pressure to catch up. To be someone by now. To finish something, prove something, accomplish something—before an imaginary clock strikes.

But here's the truth:

You don't need to chase time. You need to align with the present ripple.

That's where all transformation happens. Not in the past you mourn. Not in the future you fantasize. But in the field of now, where tension meets resolution.

There Is No "Where" in Time

Location implies space. Time implies sequence.

But both are illusions—constructs of misaligned perception.

The field doesn't operate on "before" and "after." It operates on ripple readiness. Resolution doesn't care what came before or what's supposed to come next.

"You are not behind in life. You are only ever misaligned with the present ripple."

There is no prize for speed. There is only clarity for stillness.

The Universe Isn't Waiting

You're not waiting on the universe. And it's not waiting on you.

There's no backstage manager pulling ropes, no secret timeline, no prophecy ticking.

The universe is not a play. It is an ongoing ripple dance, resolving wherever an observer becomes present enough to feel it.

And that observer...Is you.

You Are the Place Where the Cosmos Touches Meaning

Let that land.

You are not watching the world. You are where the world becomes visible.

Every wave of heat, sound, thought, light— Every insight, emotion, sensation— Every moment of clarity or confusion—They all occur where ripple tension meets you.

"The field doesn't pass by. It pulses through the portal of your alignment."

So stop looking for resolution elsewhere. Stop waiting for life to *start* when the right job, partner, idea, or answer shows up.

Be where the resolution is.

Final Words: The Mirror of Now

The present moment is the only mirror that doesn't lie.

It reflects exactly what's ready to resolve. It shows you the real. It humbles you. It aligns you.

All you have to do is return. Again and again and again.

Because...The only truth is what resolves. And it always resolves now.

The Myth of Running Out of Time

What you're fearing isn't time—it's misalignment.

"You're running out of time." "You're too late." "You missed the bus." "You should've done it by now."

These phrases sound like truth. But they're not. They're cultural hypnosis—a byproduct of linear thinking imposed on a nonlinear universe.

Time Isn't Leaving You Behind

Time, as you've been taught, is a line—with a beginning, middle, and end. You're placed somewhere along that line, told to race forward—accomplish more, faster, earlier.

But in the framework of Universal Energy Dynamics (UED), Time isn't a river carrying you downstream. It's a field of ripple tension resolving in the present.

And here's the kicker:

You cannot be *ahead* or *behind* in a system that only resolves *now*.

There is no "past due." There is no "too late." There is only misalignment with the current resolution zone.

What You're Actually Experiencing: Ripple Anxiety

When you feel like you're running out of time, you're not sensing time. You're sensing internal misalignment—a dissonance between your current ripple engagement and your desired resolution state.

In other words: It's not that life is escaping you. It's that you're not tuned in to what's already resolving in your field.

Your fear is not about a ticking clock. Your fear is about not being fully present with your ripple potential.

There Is No Universal Schedule

No tree is late to bloom. No planet misses its orbit. No ripple in the cosmic vacuum files a complaint for arriving out of turn.

So why do you?

Because human culture replaced cosmic timing with social expectation.

Degrees by 22. Marriage by 30. Success by 40. Retirement by 60.

But these are made-up intervals, mapped on a fictional linear framework.

The truth?

The universe has no deadlines. Only ripelines.

You're Not Wasting Time—You're Preparing Alignment

Every moment where "nothing seems to happen" might actually be a moment of internal field stabilization.

- Confusion is not failure. It's realignment.
- Delay is not punishment. It's tuning.
- Rest is not laziness. It's energetic integration.

You're not wasting time. You're preparing for the next ripple you are ready to resolve.

And when it aligns... you move. Not faster. Not slower. But right on time—in cosmic terms.

Final Word: Don't Run. Realign.

You are not in a race. You are in a ripple field. Nothing is leaving you behind.

Time isn't running out. Your attention is running out of the now. Bring it back. Breathe. Be where the resolution is.

Chapter 4: What Is Energy, Really?

Not a Thing, Not a Packet – It's Tension Seeking Alignment

We've built entire civilizations on a word we never truly understood—energy. We speak of storing it, transferring it, converting it. We visualize it as beams, particles, or charges in motion. But what if none of that is real? What if energy is not a thing at all—but simply the restlessness of misalignment? What if the glowing filament, the exploding star, the rising sun, and even your beating heart are all nothing more than tension resolving itself through structured ripple fields? This chapter invites you to drop the textbook definitions and meet energy where it actually lives—not as matter in motion, but as equilibrium seeking peace.

4.1 The Most Misunderstood Word in Science

Walk into any classroom, corporate office, or temple, and you'll hear the word "energy." It's everywhere—casually tossed into conversations about power, physics, performance, even prayer. We say we *store energy*, *waste energy*, *run out of energy*, *feel someone's energy*, or *harness energy from the sun*. Ask a physicist what energy is, and they'll likely reply: "The ability to do work." Ask a guru, and they might point to your chakra. Ask a layperson, and they'll show you a battery, a spark, or a shot of espresso.

But here's the truth— No one has ever touched, seen, or isolated "energy." We measure its effects, label its forms, and write equations about its conservation—but what is it, actually?

Most people picture it as:

- A glowing bolt zipping through a wire
- Heat pouring from a fire
- Light bursting from a bulb
- Calories in an apple
- Thrust from a rocket

These are not energy. These are manifestations of ripple tension resolving through specific configurations.

This book doesn't aim to mystify energy—it aims to demystify it completely. Let's drop the metaphors, the particles, the packets, the push-and-pull fairytales. Let's stop imagining energy as a *thing* and begin seeing it for what it is:

Tension in the ripple field, seeking alignment through resolution.

That's not poetry. That's the mechanics of existence. Let's begin.

4.2 Energy in Modern Science – A Useful Lie

In the modern scientific lexicon, energy is king. It powers equations, fuels experiments, explains explosions, sustains life, and drives economies. From Newton's falling apple to Einstein's nuclear epiphany, energy has been the invisible currency behind every action.

But there's one small problem. We still don't know what it is.

Science defines energy as the *capacity to do work*. That's like defining money as *the ability to buy things*—true, but hollow. Useful, but circular. Convenient, but evasive.

To make it workable, science categorized energy into neat compartments:

- Kinetic energy — the energy of motion
- Potential energy — the energy of position
- Thermal energy — the energy of temperature
- Chemical energy — the energy in bonds
- Electrical energy — the energy in charge movement
- Nuclear energy — the energy in atomic cores

But behind all of this classification is a grand assumption:

That energy is a thing—a measurable, storable, transferable *substance-like entity* moving from one form or object to another.

We speak of energy transfer as if it were a parcel in the mail. We calculate energy conservation as if we're tracking coins in a ledger. But when you ask the question that cuts too deep for comfort—What is actually being transferred? You're met with silence. Or metaphors. Or mathematical tricks that work but don't reveal.

The truth is: science has made energy into a symbol of cause and effect. A placeholder for "something happened here" or "something changed there." It's not wrong—it's incredibly useful. But it's not real in the way we imagine it. It has no shape, no substance, no ontology of its own.

The lie, if we must call it that, is not malicious. It's functional. It helped us build machines, launch rockets, light up cities, and treat disease. But now, it's time to go deeper. To ask: *What lies beneath the symbol? What is the field really doing when we say "energy was transferred"?*

The answer is not in a wire or a wave. It's in tension, and how the universe resolves it.

4.3 UED Perspective – Energy as Ripple Tension

What if we told you that energy doesn't move—but imbalance does? What if every machine, every body, every explosion, and every emotion was not powered by energy—but by misalignment resolving itself?

Welcome to the UED perspective:

Energy is not a thing. It is a condition. A state of unresolved ripple tension in the structured vacuum field.

In traditional science, energy is a carrier of change. In Universal Energy Dynamics (UED), energy is the consequence of field imbalance. It does not exist as a stored packet or flowing entity—it emerges as ripple tension attempts to realign with its surrounding equilibrium.

Let's break it down:

- When you lift a rock and hold it in the air, traditional science says the rock has *gravitational potential energy*.
- UED says: the rock is in a ripple misalignment zone—its position disrupts the ripple field's preferred equilibrium.
- Release the rock, and what happens? Not energy transfer—but ripple resolution. The tension dissolves, aligning back with the structured vacuum, and the "motion" is merely the visible consequence of that return to stillness.

This shift changes everything.

In UED:

- A charged battery is not storing energy—it's storing ripple strain in a structured configuration.
- A wound spring is not holding energy—it's resisting ripple alignment.
- A human body is not fueled by calories—it is navigating ripple pressures across biochemical fields, metabolizing imbalance toward harmony.

That's why the moment tension resolves—heat, motion, light, and experience *appear*. But these are after-effects, not cargo being delivered.

You don't carry energy. You carry imbalance. And when that imbalance finds a path to resolution— Reality happens.

UED doesn't reject energy. It redefines it. Not as a unit, not as a wave, not as a particle— But as the language of the field trying to become whole again.

4.4 There Are No Packets – The Photon Myth

For over a century, we've been taught to think of light and energy in terms of particles—tiny packets flying through space, delivering power, illumination, or heat. The photon, in particular, has become the poster child for this idea:

A "massless particle" that moves at the speed of light and carries energy in quantized bundles.

Elegant? Yes. Mathematically useful? Absolutely. But real? Not quite.

In the UED framework, the photon is not a thing. It is a ripple event—a *localized resolution* of tension within the structured vacuum field. It doesn't fly from the Sun to your eye. It doesn't bounce off mirrors or pass through glass. It appears where ripple alignment happens. And only there.

Let's take sunlight as an example:

- Conventional view: photons are emitted from the Sun, travel 150 million kilometers through space, hit Earth, and interact with surfaces and eyes.
- UED view: the Sun's ripple field interacts with Earth's ripple field, creating tension across distance. That tension gets locally resolved at the atmospheric boundary—and *that* is what we see as sunlight.

Nothing traveled. Nothing flew. What changed was the field configuration—a resolution of ripple mismatch that becomes light when aligned with an observer's perceptual system.

This principle doesn't apply just to light:

- Electricity doesn't flow through wires—it's sequential ripple alignment along conductive paths.
- Radiation doesn't shoot out—it's field reconfiguration spreading through structured space.
- Even so-called heat transfer is nothing more than ripple tension diffusing to surrounding field zones.

The idea of a "packet" is a mental shortcut. A metaphor. A placeholder.

But packets don't exist. Patterns do.

Every observed "quantum" of energy is not a traveling entity, but a field-based resolution signature—a unique ripple pattern triggered by boundary conditions.

The photon, in this view, is not a messenger. It is a mirage caused by our need to materialize experience.

This reframing pulls the rug from under much of quantum orthodoxy—but it doesn't leave you in the dark. Instead, it reveals a deeper light:

One that never traveled—only aligned.

4.5 Energy Conversion = Ripple Reconfiguration

We've been taught that machines convert energy from one form to another. Textbooks say:

- Electricity turns into heat inside a toaster
- Fuel turns into motion in an engine

- Sound turns into electrical signals in a microphone

That's the traditional story. But what's really happening?

Let's reframe the entire concept through Universal Energy Dynamics (UED):

No energy flows. No substance is transferred. There is only ripple tension, reconfigured.

Every so-called "conversion" is actually a field re-alignment event. Each machine or system is just a patterned interface, built to resolve ripple tension in a specific way.

Toaster Example:

- You apply electric tension (ripple misalignment across a conductive path).
- The coiled wires offer structured resistance.
- This tension resolves as localized thermal ripple excitation—what we feel as heat.

Nothing "flowed." The toaster didn't "consume" electricity—it aligned ripple strain into heat zones.

Combustion Engine:

- Fuel holds high ripple imbalance in chemical form.
- Ignition breaks molecular alignment, releasing tension.
- Pistons channel this resolution into mechanical ripple motion (vibration, rotation, thrust).

The engine is not a converter. It's a structured ripple realigner, designed to take one type of field tension and reformat it into another.

Microphone:

- Sound enters as air-pressure ripple tension.

- The diaphragm vibrates in sync, realigning that tension into an electromagnetic pattern.
- The output signal is not a "converted sound," but a new ripple expression—tuned to be resolved later by a speaker or a brain.

This is the great mechanical illusion of modern science:

We don't have machines that convert energy. We have machines that participate in ripple choreography—designed for sequential resolution.

What looks like "energy moving" is just field conditions shifting across well-defined ripple paths.

And you, the observer, interpret these ripple sequences as "function," "motion," or "output."

So what is a machine, really?

A ripple realigner. A temporary sculptor of tension. A tool to reshape unresolved conditions into meaningful outcomes.

Not by pushing, pulling, or flowing— But by letting ripple fields resolve themselves through intelligent design.

4.6 Stored Energy? No—Stored Tension

We often say:

- "The battery stores energy."
- "The spring holds energy."
- "Chemical bonds contain energy."

But pause for a moment: What does it really mean to "store" energy? Is there a glowing substance waiting to be released? A packet tucked away for later use?

UED answers boldly:

Nothing is stored. What we call "stored energy" is simply ripple tension held in an unresolved configuration.

Take a Battery:

It doesn't hold electricity like a jar holds water. It maintains a structured misalignment in the ripple field between its terminals.

- Positive and negative ends = ripple poles.
- Electrochemical reactions maintain a persistent tension.
- When connected in a circuit, this ripple strain resolves, and the appearance of "current" is just field realignment in motion.

The battery didn't *release* anything. It simply allowed tension to express itself through alignment.

Take a Spring:

When you compress or stretch a spring, you're distorting its equilibrium. The metal's internal lattice becomes geometrically misaligned.

- This doesn't "load" the spring with energy.
- It encodes ripple resistance—a preference to return to rest.

Let go, and the spring reverts not because of energy flow, but because ripple tension collapses back to stable alignment.

Take a Chemical Bond:

In molecules, atoms share electrons in tightly balanced ripple arrangements. But certain bonds—like those in fuel or food—are unstable alignments, ready to reconfigure.

- The tension in these bonds is not energy stored in a box.
- It is an unresolved ripple pattern—primed for collapse.
- Combustion, digestion, metabolism—these are just sequences of re-alignment, playing out in structured stages.

So let's set the record straight:

"Potential energy" is not hidden fuel. It is ripple pressure waiting for release—a condition, not a commodity.

And the moment the system finds a resolution path, what we call "energy release" is simply a rapid, cascading restoration of alignment.

Implication: The world isn't running on stored energy. It's running on strategically delayed harmony. Systems work because they are delicately held out of balance, then allowed to settle in ways that seem functional.

Or in simpler terms:

What we call potential is simply unexpressed ripple pressure.

4.7 Heat, Motion, Light – All the Same Game

When we say something *has* energy, we usually mean it in one of many fragmented ways:

- It's hot (heat energy)
- It's moving (kinetic energy)
- It's glowing (radiant energy)

Science has categorized these into "forms of energy," implying that energy comes in different flavors that can be shuffled, transformed, stored, or transferred like cards in a magician's deck.

But from the lens of Universal Energy Dynamics (UED), this illusion collapses.

All Forms Are Just Faces of One Thing: Ripple Resolution

There is no fundamental difference between heat, motion, or light. They are all the perceptible outcomes of ripple tension resolving at different scales and through different interfaces.

Let's break it down:

Heat

What we interpret as heat is not a substance traveling from hot to cold.

- It is the local resolution of misaligned ripple tensions in the thermal bandwidth.
- The "hotter" something is, the more ripple agitation it exhibits within a field.
- Your skin doesn't feel heat because it *receives* anything—but because it resolves ripple misalignment into sensation.

Motion

Motion is not an object "going somewhere" in space.

- It is sequential realignment of ripple tension along a chosen path.
- When a car moves, what's really happening is that its body is being progressively re-anchored in the SVE field by a cascade of ripple shifts.
- What we call acceleration or velocity is a rate of ripple reconfiguration—not spatial migration.

Light

Light isn't a beam traveling from one place to another.

- It is a visible outcome of ripple resolution between the radiating field and a receptive observer (eye, sensor, material).
- There's no photon flying through space. Just ripple tension resolving at precise optical bandwidths.

The Critical Shift

So instead of asking:

- *How is heat different from motion?*
- *How does light "carry" energy?*

Ask instead:

- *What ripple bandwidth is being resolved?*
- *Where is the tension highest—and how is it being expressed?*

Because all that exists is tension and resolution.
The UED Insight:
Energy is not in things. It is between alignments.

- A hot iron, a sprinting runner, a glowing bulb—they don't "contain" energy.
- They are just zones of active ripple reconfiguration.
- Once alignment completes, the "energy" seems to disappear—because the tension is gone, not because a substance was used up.

Sensory Trickery

Every experience—heat on the skin, light in the sky, vibration in the chest—is not an input, but a resolution interpreted by the observer.

- Sight = optical resolution
- Sound = pressure resolution
- Touch = mechanical resolution
- Warmth = thermal resolution

Change the resolution parameters, and the "form of energy" changes too.

In truth, there is no difference between the warmth of the sun, the movement of a leaf, or the flash of lightning. Only the interface—and the resolution path—makes them appear different.

They're all the same game:

Tension resolving itself through the fields we call "matter."

4.8 Closing Reflection – Seek the Tension, Not the Flow

We've been trained—almost hypnotized—to look for "energy" in everything.

Where's the fuel? Where's the power source? Where's the energy *going*?

But this entire way of thinking belongs to a false paradigm—a world that imagines energy as a flowing commodity.

Modern science, education, engineering—even spirituality—has built entire empires atop this belief. "Energy flows," they say, "from source to sink." But stop. Look again.

Nothing flows. Everything resolves.

The Hidden Engine of the Universe

The true driver of all motion, sensation, transformation, and existence... is not *energy transfer*. It is ripple tension seeking alignment—the universe trying to restabilize itself, field by field, layer by layer, interaction by interaction.

You don't burn fuel. You uncoil a held misalignment and reconfigure it through ripple interfaces.

The sun doesn't "emit" light. It's a cosmic tension zone, resolved by fields like Earth's atmosphere—and ultimately, your eye.

The battery doesn't "store" energy. It holds unresolved ripple tension, waiting for an aligned interface (like a circuit) to release it.

Shift the Question

Don't ask: "Where is the energy going?"

Ask instead:

"What is trying to resolve here?" "What tension is present?" "How is the field attempting balance?"

Every machine, every life-form, every experience—from a running engine to a beating heart—is a ripple resolution in progress.

You Are Not Consuming — You Are Participating

You are not a consumer of energy.

You are an interface, a decoder, a co-resolver of universal tensions. When you breathe, move, think, feel—it is not because some "energy" flowed into you. It is because your existence aligns momentarily with tension fields being resolved.

This is not mysticism. This is mechanics of participation.

Open Your Eyes: Everything Around You is a Resolution Field

Look around now:

- A ceiling fan spinning
- A cup of hot tea cooling
- A phone glowing with light
- Your own chest rising and falling

None of this is powered by anything *moving through space.* It is all local ripple resolution, happening at the interface between structure and tension.

Final Thought:

"You are not consuming energy. You are participating in ripple resolution."

And the more aware you become of this, the more precise, aligned, and intelligent your interaction with the universe becomes.

So don't chase the current.

Seek the tension. Resolve the ripple. And watch your entire world transform.

Interlude: Before the Mirror Blinks

You've seen the light bend. You've felt the heat. You've heard the sound. You've called it real—without ever asking what made it so. But now, something inside you knows: It wasn't the sun that glowed. It wasn't the air that moved. It wasn't even the world that changed.

It was you—aligning.

Everything you thought you knew about motion, energy, sensation, even time—was a mirage dancing in the field. A ripple resolving itself where *you* happened to be tuned.

There are no things. No photons. No waves. No particles. Only patterns. Only pressure zones. Only tension and stillness—writing their play upon your perception.

And the stage? It was always still. What does that make you? A machine? An animal? A conscious bag of biology in motion?

No. It makes you the decoder. The aligner. The one who sees the ripple and gives it meaning.

- You thought you were walking through life. But life was resolving through you.
- You thought you were receiving energy. But you were meeting imbalance—and shaping the resolution.
- You thought reality happened to you. But now you know—it happens with you.

So pause here. Breathe. And prepare.

Part II: Life Inside the Ripple

Every cell knows what you forgot.

Chapter 5: Every Atom Is Aligned

Structured Vacuum Energy (SVE) and Ripple Architecture

You've been taught that atoms are tiny balls—building blocks that form the solid world. But what if the atom is not a thing at all? What if it's a stillness in disguise—a beautifully held tension in a sea of structured vacuum? In this chapter, we pull back the veil on "matter" and reveal the ripple logic beneath its apparent solidity. There are no objects here—only agreements. The atom exists not because something is *there*, but because something is *aligned*. Structured Vacuum Energy—SVE—is not emptiness. It is the hidden architecture of reality, a silent field that holds everything in delicate, dynamic balance. And every atom you've ever touched... is simply a ripple that never collapsed.

5.1 There Are No Solid Things

You knock on a table. It feels firm. You pick up a stone. It resists your grip. You touch your body—it pushes back with presence. And so, you believe: this is solid, this is real, this is matter.

But pause.

What if solidity is the most convincing illusion of all?

What if that table, that stone, your very body... are not "things," but *resolutions*? What if everything you call solid is simply a ripple caught in perfect balance?

We've been trained to equate resistance with substance—if it pushes back, it must be there. But resistance can arise from *tension*, not from mass. From *structure*, not from stuff. Think of a tightly stretched trampoline. You can't pass through it, not because it's a wall, but because of how its fabric holds tension. Now apply that metaphor to the universe itself.

Everything you've ever seen, touched, or thought of as "material" is actually a dance of alignments—a standing ripple formation within the Structured Vacuum. The atom is not a ball. It is a ripple equilibrium. The nucleus is not a lump. It's a stabilized storm of field tension.

This changes everything.

Solidity is not what you're holding—it's what the universe is *holding in place* through tension dynamics. What appears as hardness is just local alignment dense enough to push back against your field.

You are not surrounded by objects. You are swimming in arrangements.

And once you see this—not with your eyes, but with understanding—you will never see "solid things" the same way again.

5.2 What Is Structured Vacuum Energy (SVE)?

We've been told space is empty. A cold nothingness. A backdrop for the real action—matter, light, energy. But what if we've misunderstood the stage entirely?

The Structured Vacuum Energy, or SVE, is not an empty void. It is the hidden fabric that holds it all—still, silent, yet pregnant with potential. It is not passive, but poised. Not chaotic, but exquisitely organized.

SVE is not nothing—it's everything that appears as nothing.

Imagine a sheet pulled tightly in every direction. To the eye, it seems still. But try to press it. You'll feel the tension. You'll see ripples emerge. That is SVE—a hyper-stable, pre-configured field that supports ripple formation without itself ever "moving." It's not material, but it enables matter. It doesn't flow, but it allows all resolution.

Unlike classical vacuum—which physics once modeled as an absence of particles or energy—SVE is structured presence. It is a tension lattice, a ripple-friendly geometry that remains in perfect equilibrium until disturbed.

And when it is disturbed? That's when reality begins.

Waves don't travel through space. They restructure SVE. Particles don't move—they appear as stable ripple formations within this structured field. Light, gravity, mass, force—all emerge as interactions with this pre-existing ripple-capable substrate.

In Universal Energy Dynamics (UED), the SVE is the default state of the universe—a zero-action field of infinite potential alignment. Not neutral like empty space, but zeroed like a guitar string before it plays a note. It holds no observable "stuff," but it holds the rules for every "thing" to come.

Every atom is a local disruption in SVE—a standing ripple, frozen in equilibrium. Every event, every perception, is a temporary misalignment resolving back into that silent structure.

So the next time you imagine "vacuum," don't picture a void.

Picture a perfectly still sea that knows exactly how to ripple—when you show up.

5.3 The Atom Is Not a Thing, It's a Ripple Configuration

We were taught that atoms are tiny solar systems. A dense nucleus of protons and neutrons at the center. Electrons orbiting like planets.

A neat, intuitive picture—simple enough for schoolbooks, useful enough for calculations.

But fundamentally wrong.

In the Universal Energy Dynamics (UED) framework, the atom is not a "thing" composed of smaller "things." It is a ripple configuration—a stable standing pattern in the Structured Vacuum Energy (SVE) field. Not orbiting particles. Not solid spheres. But zones of concentrated ripple tension, held in dynamic balance.

Let's begin at the center:

- The nucleus is not a "ball of mass." It is a high-tension ripple knot—a tight, compressed configuration of SVE tension gradients. What we call "protons" and "neutrons" are nothing but harmonic pressure concentrations—interference zones that have temporarily stabilized.
- These ripple knots are not glued together by mysterious "strong nuclear forces." They are locked into equilibrium by the balancing of their ripple tensions, much like tuning forks resonating at coherent frequencies when placed side by side.

Now let's go to the edge:

- Electrons are not orbiting planets. They are not particles in the classical sense. Instead, each "electron" is a field-compatible oscillation—a rhythmic, boundary-specific ripple that maintains a quantized resonance with the core structure.
- These "shells" or "orbitals" are not paths or places—they are stable alignment zones in the surrounding SVE lattice. The electron appears only when and where resolution occurs—never in between. That's why it behaves like both

wave and particle. Because it is neither—it is a resolvable ripple potential, not a permanent thing.

Why does it appear stable then? Why doesn't the atom collapse or explode?

Because the SVE is tuned to hold such configurations. The ripple tensions of each atomic configuration represent a minimum energy alignment—not because something is "trapped," but because it has nowhere better to resolve.

The atom is not an object—it is a stabilized field story. It's the quiet agreement of ripple tensions that decided to play a consistent note.

So next time someone asks you what an atom is, don't say "a building block." Say this instead:

"It's a ripple frozen in song, held perfectly by the silent structure of the universe."

5.4 Orbitals = Alignment Patterns

When we first encounter the structure of the atom in school, we're shown electrons orbiting in neat rings—like planets. Later, we're told that those rings are just metaphors, replaced by "clouds" of probability. Still, most people walk away imagining little balls zooming around a nucleus.

But in UED, we discard both metaphors. There are no rings. No orbits. No clouds.

Instead, orbitals are alignment patterns—zones of ripple tension compatibility between the atomic core and the surrounding Structured Vacuum Energy (SVE). These are not locations where "an electron is," but regions where resolution can happen. In short:

Orbitals are not maps of movement. They are maps of possibility.

The s, p, d, f Orbitals = Field Resonance Modes

Each orbital type—spherical (s), dumbbell-shaped (p), cloverleaf (d), complex (f)—is not a shape of an object, but a shape of alignment.

Imagine plucking a taut membrane. The ripples that form depend on the shape, tension, and boundary conditions. Atoms work the same way.

- The s-orbital is a fully symmetric pressure zone—equal tension in all directions.
- The p-orbital introduces dual lobes—localized alignment on either side of a nodal plane.
- d and f orbitals are higher-harmonic tension arrangements—complex zones of field stability where ripple patterns fold, twist, and knot in intricate geometry.

These shapes emerge not because something is moving in those directions, but because the ripple resonance between nucleus and SVE naturally stabilizes in those configurations.

Quantum Numbers = Ripple Harmonics

In traditional quantum mechanics, orbitals are governed by four quantum numbers—n, l, m, and s. But what are these really?

In UED, each of these is not a quantity for a particle—it's a harmonic descriptor of ripple geometry:

- n (principal) = tension layer depth (number of major ripple zones)
- l (angular) = angular symmetry (type of nodal alignment)
- m (magnetic) = orientation within external field ripple (fine-tuned axis)
- s (spin) = micro-ripple orientation within the stable tension pocket

They do not describe where a particle is, but how the ripple field is structured in that local SVE environment.

Electron "Clouds" = Resolution Probability, Not Presence

What we call the "electron cloud" is a map of ripple resolution likelihood. It is not the smear of a fast-moving particle. It is a field of readiness—zones where, if alignment is probed (say, by light or measurement), resolution might manifest an "electron-like event."

Until then, nothing is flying around. There is only potential.

To summarize:

Orbitals are not electron paths. They are patterns of tension that hold ripple possibilities. Quantum numbers are just the musical notation of this silent geometry.

And thus, the atom doesn't contain moving parts. It contains ripple music, frozen into standing harmony with the universe.

5.5 Why Stability Means Alignment, Not Bonding

In traditional chemistry, atomic stability is taught as the result of bonding—an act of electron sharing, transferring, or overlapping. Covalent, ionic, metallic, hydrogen bonds—each treated as a kind of handshake between atoms. But this view, while practical for lab equations, misses the deeper field-based truth.

In Universal Energy Dynamics (UED), we must go beneath the particles and into the field: There is no "bond." There is only balance.

What Makes an Atom Stable?

An atom is said to be "stable" when it doesn't easily react or transform. But what does this actually mean?

In UED, an atom is stable when its internal ripple tensions—between the nucleus, its orbital architecture, and the surrounding SVE—settle into a local equilibrium. This is not static, but dynamically balanced: like a tightrope walker who is always moving, but never falling.

Stability, then, is not about possession of electrons, but about cancellation of internal ripple tensions through geometric symmetry and field resonance.

What, Then, Is a Bond?

A chemical bond is not an object. It is a ripple interlock—a localized zone where multiple atoms' ripple patterns find shared resolution.

- A covalent bond is not a shared electron. It is a mutual alignment zone where the SVE ripple tensions of both atoms find lower energy together than they did apart.
- An ionic bond is not a transfer of electrons from one to another. It is a field polarity compensation, where charge differential across ripple structures resolves tension imbalance.
- A hydrogen bond is a subtle field coupling, where a nearby molecule's ripple geometry offers partial support to another's unstable region.

In short:

Bonding is not sharing particles. It is co-resonance in a structured field.

Molecules = Complex Ripple Harmony

When multiple atoms bond, the result is a molecule—not a cluster of stuck-together balls, but a standing wave architecture, where ripple pressures interweave, cancel, and reinforce into a meta-stable pattern.

The molecule is not a structure of objects. It is a ripple solution.

Every molecular vibration, every conformational change, every chemical behavior—is a readjustment in field tension, a dance of ripple equilibrium across participating atoms and the medium of SVE.

The Takeaway:

- Stability = minimum ripple misalignment
- Bonding = alignment interlock, not particle exchange
- Molecule = harmonic ripple system, not a bag of atoms

So next time you see a molecule diagram, don't picture balls and sticks. Picture a ripple lattice in song, holding tension in just the right geometry to stay alive.

"What you call a bond is simply the quiet between opposing tensions."

5.6 No Atom Is Alone

Even in the cleanest vacuum chamber, surrounded by layers of isolation, an atom is never truly isolated. That's not just a philosophical sentiment—it's a physical truth. In Universal Energy Dynamics (UED), every atom exists as a localized resolution pattern in the Structured Vacuum Energy (SVE) field, and that field, by nature, is continuous.

There are no walls in SVE. Only gradients of tension.

Ripple Fields Extend Beyond Boundaries

An atom's structure is not confined to a shell. Its ripple field—just like gravity, magnetism, or light—extends outward, blending and interacting with its environment. The further from the center, the weaker the tension influence—but never zero.

In UED terms:

- An atom is not a "point object."
- It is a ripple configuration with a zone of influence, shaped by how its internal tensions are resolving against the

background field.

This ripple influence allows subtle interactions even between atoms that appear "non-bonded" or "far apart." This explains why:

- Gas molecules repel or attract slightly without bonding.
- Atoms in solid lattices adjust depending on nearby substitutions.
- Spectral lines shift slightly in different environmental contexts.

Field Coupling and Environmental Shifts

Place a noble gas atom like argon in a vacuum—its behavior is predictable and "inert." Now place it in a dense atmosphere or plasma. Suddenly, that "inertness" bends. The energy levels shift, polarizability changes, and ionization thresholds adjust.

Why?

Because the ripple field of argon is now resolving against different tensions in the surrounding SVE. That environment—though not forming chemical bonds—alters the field balance. The atom must realign subtly to maintain equilibrium.

This is field coupling—a kind of tension whisper between nearby configurations.

Alignment Memory: Atoms Remember Their Field History

UED introduces a fascinating concept: alignment memory.

Even after leaving a region, an atom's field may retain a slight bias—a "memory" of the ripple alignment it experienced. Like a tuning fork that still hums faintly after the sound stops, atoms carry subtle imprints of their previous field entanglements.

This memory can influence:

- Spectral behavior
- Reactivity patterns

- Structural preferences in complex systems

It's why biological molecules fold differently in different environments, why materials behave differently under stress, and why no two "identical" atoms ever act identically in the real world.

Even Inert Atoms Are Contextual Beings

In classical chemistry, noble gases are painted as loners—non-reactive, aloof, complete.

But UED says otherwise:

- Their electron shells may be stable,
- But their ripple fields are still active.

Place them near charged plates, inside magnetic traps, or around dense mass—they will respond. Not with dramatic reactions, but with quiet realignments, shifts in field balance, subtle behavioral changes.

Because no atom is sealed off from the field.

Summary

- Atoms are not islands—they are ripple nodes in a continuous sea.
- "Inert" doesn't mean immune—it means temporarily aligned.
- Field context, not chemical identity, dictates behavior.

"An atom does not end at its shell. It ends where alignment fades."

5.7 The Soul of the Atom?

Let us ask a question that mainstream science avoids—not because it's unscientific, but because it's uncomfortable:

Why does the atom stay together? What holds this delicate ripple of tension in such beautiful, sustained alignment? What gives it the intent to remain stable?

Beyond Binding: The Mystery of Continuity

Physics offers us formal equations: electromagnetic forces, quantized energy levels, conservation laws. Chemistry gives us shell models and bonding diagrams. All useful. All partial.

But step back. Look again.

Here's a structure—an atom—that:

- Resists collapse,
- Adapts to external field shifts,
- Realigns when perturbed,
- And recovers equilibrium after excitation.

This isn't just structure. This is behavior. This is intelligent tension—a field striving for self-consistency across change.

Alignment Is the Seed of Intelligence

In UED, intelligence isn't stored in a brain. It is a consequence of coherent alignment within structured tension.

- A ripple that can sustain its structure against noise—has memory.
- A field that can adapt its resonance without losing form—has self-correction.
- A node that can interact without losing identity—has presence.

The atom does all of these.

Which means: The atom is not intelligent like a mind. But it is not dumb like a rock either.

It is alive with pattern. Not sentient—but resonant with a will to persist.

This "will" is not an emotion or desire. It's the ripple's bias toward harmony—the same universal principle that forms galaxies, proteins, and poems.

SVE + Resolution = The First Pulse of Intelligence

Structured Vacuum Energy is the canvas. Ripple tension is the motion. But resolution—that act of returning to balance—is where intelligence begins.

The atom's form is a solution to the field's question: "How can tension hold itself in balance?"

This solution repeats, adapts, evolves, and multiplies. It builds hydrogen. Then carbon. Then DNA. Then humans.

And it all begins with a single aligned ripple.

Introducing Universal Intelligence (Softly)

So is there a "soul" in the atom?

Not in the religious or mystical sense. But in the UED framework: Yes, there is a trace of universal intelligence embedded in its design. Not awareness. But alignment.

Not a watcher. But a rhythm that knows how to hold itself together.

Atoms don't think. But they remember tension. They don't choose. But they respond with precision. They don't live. But they don't die either.

They are the first stable whispers of something deeper: A universe not made of things, but of intelligent resolution.

"What if the soul is not a spark of something else—but the echo of balance itself?"

5.8 Closing Reflection – The Aligned Foundations of Reality

What if we told you— That you've never touched a "thing" in your life?

That everything you've ever held, sat on, eaten, or loved...was never made of matter as you were taught— but of field tensions resolved into form?

Matter: The Grand Illusion of Solidity

Let's rewind. What is a rock? Not a pile of particles. But a dense lattice of aligned ripple tension.

What is your hand? Not flesh and bone in isolation. But an orchestration of atoms holding their shape in an invisible ripple network.

What we call "matter" is simply what resolves consistently enough to appear solid. Nothing is actually solid. Everything is stable.

From Particle to Pattern

The old worldview told us:

- "The universe is made of particles."
- "They bounce, collide, exchange force, and make up stuff."

But UED clears the fog:

- There are no particles.
- There are only patterns of alignment.
- What looks like a particle is simply a stable solution to ripple tension.

Even subatomic "objects" like quarks or electrons aren't objects—they're states. Energy states. Field alignment states. Not billiard balls. Not dots.

Just coherent ripple zones holding their shape long enough to mean something.

The Memory of Ripple Resolution

Each atom is a memory. Not of events, but of alignment.

It remembers how to hold form. It remembers how to respond when heat comes, when a photon strikes, when a molecule approaches.

Atoms "behave" not because they are intelligent—but because they are resolved ripple behaviors stored in SVE. Their structure is a contract between the field and the tension— A treaty of balance, constantly maintained, rarely broken.

Matter Is Agreement, Not Object

This is perhaps the most radical shift:

"Matter is not made. It is resolved."

Reality is not a construction of pieces. It is a continuous unfolding of agreements—between ripple and field, between alignment and pressure, between now and next.

The universe doesn't build with bricks. It breathes with balance. It doesn't operate on cause and effect—it harmonizes through equilibrium seeking.

And the atom—our so-called building block—isn't a block at all. It is a beautiful compromise between structure and possibility.

"There are no particles—only places where tension remembers how to hold itself."

"The atom is not a thing. It is a decision the universe made, and keeps making—perfectly, constantly, silently."

Chapter 6: The Engineering of Living Things

How Insects, Birds, Plants Manipulate Ripple Resolution

Before humans built machines, wrote formulas, or drafted blueprints, nature was already a flawless engineer. Without language, logic, or laboratories, living organisms—bees, birds, trees, fungi—were orchestrating ripple alignment with a finesse that humbles our most sophisticated inventions. They don't *calculate* alignment—they **live it**. Every flap of a wing, every twist of a vine, every curve of a hive is a direct expression of field intelligence responding to local tension. In this chapter, we step away from human ego and into the quiet brilliance of life itself—a realm where ripple resolution isn't theorized, it's embodied. Welcome to the living blueprint of the universe.

6.1 Nature as the True Engineer

We've long believed that engineering is a human domain—blueprints, machines, mathematics. But before blueprints, there were beehives. Before algorithms, there were ant trails. And before propulsion, birds launched themselves into the sky. What if nature has always been the superior engineer, not because it knows equations, but because it embodies resolution?

Let's challenge the tired narrative of evolution as a slow game of trial and error. What if the rise of wings, sonar, photosynthesis, or camouflage wasn't random mutation filtered by survival, but rather a deeper intelligence—ripple intelligence—emerging through alignment?

Every living thing operates with field sensitivity. A plant doesn't grow leaves randomly—it aligns its surface to optimize light resolution. A bird doesn't fly by effort—it glides by surrendering to field gradients. An insect doesn't wander—it traces ripple gradients like a living antenna.

The most astonishing truth? None of them need a "brain" to do this. Their structure itself is tuned. They don't calculate—they resonate. Life doesn't fight the field. It flows with it, aligns with it, and engineers solutions by dissolving resistance.

Nature doesn't ask how to survive. It aligns—and survival follows.

6.2 Insects: Micro Masters of Field Response

Insects are often dismissed as primitive or simple—but in truth, they are precision instruments of ripple intelligence. Their bodies, movements, and behaviors are not random nor purely instinctive. They are deeply aligned with their environments, operating as miniature engineers of the field. These beings do not manipulate the world with tools—they manipulate tension, temperature, geometry, and vibration, often with more finesse than any human technology.

Bees: Hexagons of Harmony

The hexagonal structure of a beehive isn't just a geometric fluke—it is a pressure-optimized ripple form. Among all tessellating shapes, the hexagon encloses the most space with the least material and stress. But more than that, it naturally distributes ripple tension evenly through the hive matrix. Bees don't "calculate" this—they align to the field and express geometry that resolves construction stress efficiently. The comb is not built—it emerges from tension equilibrium.

Ants: Decision Gradients Without Brains

Ant colonies demonstrate a kind of collective field intelligence. No central controller directs them, yet they navigate complex terrain, build intricate tunnels, and coordinate vast colonies with precision. How? Through local ripple cues: pheromones act as alignment triggers, creating gradient paths of decision. One ant's signal nudges another's resolution field. Over time, these fields couple, forming a collective intelligence that moves like a single organism—despite no individual "knowing" the plan.

Dragonflies: Aerodynamics via Ripple Mastery

Dragonflies can hover, dart, pivot, and glide with uncanny accuracy. Their secret lies in the real-time modulation of wing tension. Every tiny shift in pressure is registered, and the insect's structure adjusts accordingly. It's not muscle strength that allows this—it's field-sensitive ripple feedback. Their wings act as live antennae for air currents—resolving tension mid-flight, allowing aerial acrobatics no drone can match.

6.3 Birds: Flight by Field Geometry

Birds are not merely creatures of sky—they are engineers of alignment, navigating not through brute force, but through an exquisite relationship with the surrounding ripple fields. Their biology is not just suited for flight—it is designed for field harmony, enabling them to interact with gravity, magnetism, and atmospheric tension in ways that modern aerospace engineering still struggles to mimic.

Feather Design = Aerodynamic Ripple Diffusion

Feathers are not just light and beautiful—they are structured for field management. Each feather's micro-barbs and flexible shaft don't simply slice through air—they diffuse aerodynamic ripple tension across the bird's body, creating dynamic balance between lift, drag, and thrust. In high-speed dives or effortless glides, it's not just muscle—it's feathered tension intelligence. Each twitch of a wing tip shifts ripple alignment, creating flight through precision, not resistance.

Migration = Navigation via Ripple Gradients

Birds migrate thousands of kilometers across oceans and continents, often without visual cues or prior experience. How? Earth's magnetic field isn't just a compass—it's a ripple gradient that birds tune into. Their internal magneto-receptors don't "detect direction"—they align with ripple flow, sensing tension patterns in the planet's field architecture. Migration, then, is not a route to follow—it's a field to feel. The Earth whispers alignment, and they move.

Birdsong = Frequency-Tuned Field Imprinting

When birds sing, they're not just making sound—they are imprinting the ripple field with signature frequencies. Each chirp, call, or melody is a vibration tuned for territorial resonance, mating alignment, or flock cohesion. These are not mere audio signals—they are ripple-stamped intentions, sensed by other birds in the surrounding field. A song is not heard—it is felt—as alignment or disharmony.

Flight = Not Defiance, but Dance

We often think of flight as "defying gravity," but birds show us a deeper truth: flight is field alignment. They don't resist gravity—they collaborate with it. By entering the ripple structure of air and adjusting posture, feather spread, and angle, they surf the unseen tension—riding equilibrium like a wave. The sky is not an obstacle—it is a responsive medium. And birds? They are its fluent dancers.

6.4 Plants: Stillness That Engineers

They don't move, yet they adapt. They don't speak, yet they respond. Plants are not passive participants in nature—they are masters of stillness-based engineering. Beneath their calm surface lies an extraordinary ability to read, resolve, and reconfigure ripple tensions—not by motion, but by sustained alignment with their environment. What looks like slow growth is actually a deliberate architectural dialogue with the field.

Sunflower Alignment (Heliotropism) = Ripple-Following for Max Absorption

Sunflowers don't "track the sun" because they love light—they do it because they're resolving thermal and electromagnetic ripple fields for maximal intake. Each stem subtly bends—not randomly, but in response to field gradient shifts from morning to evening. This isn't photosynthesis alone—it's ripple choreography. The sunflower's face is not just turning toward light—it is aligning with peak resolution tension for optimal energy conversion.

Leaf Orientation = Thermal and Moisture Balance via Ripple Regulation

Leaves are finely tuned surfaces—living sensors of temperature, humidity, and solar flux. Their orientation adjusts to minimize ripple pressure imbalances, regulating moisture loss (transpiration) and thermal load. A vertical leaf might be avoiding excess solar heat; a horizontal one might be maximizing ripple absorption. Every tilt is a ripple decision, optimizing alignment for survival—not through thought, but through encoded intelligence.

Roots = Subsurface Ripple Reading and Re-Routing

Roots don't just grow blindly into the soil. They sense moisture gradients, mineral fields, and structural resistance—all through ripple pressure differentials below ground. Root tips adjust direction in real-time, seeking lowest resistance paths and nutrient-rich alignments. Some roots even avoid underground obstacles long before contact—not by sight, but by ripple sensing. Their intelligence is subterranean—but no less precise.

Vines = Active Response to Tension Imbalance in Nearby Support Fields

Climbing plants like vines showcase one of the most dynamic ripple strategies in the plant world. They actively search for supportive field tensions—walls, trees, or poles—and extend tendrils toward them. When contact is sensed, the plant doesn't just hold—it twists, spirals, and locks in patterns that maximize mechanical stability and energetic efficiency. This is not random wrapping—it's field-based architecture, responding to local tension zones.

In essence, plants show that stillness is not inactivity. It is complete awareness. Every growth, every bloom, every turn of a leaf is a response to ripple. Plants don't fight their environment—they align with it. They engineer without ego, optimize without movement, and teach us that true intelligence can exist in perfect silence.

6.5 Symbiosis = Shared Alignment Contract

In nature, survival is not a solo act. It's a collaborative play of ripple-aware beings, each tuned to their unique bandwidth, but often entangled in mutual alignment strategies. Symbiosis is not merely "coexistence." It is a contract of shared ripple resolution—where different lifeforms enter a dynamic agreement to stabilize field tensions together.

Mycorrhizal Fungi and Plant Roots: Co-Resolved Nutrient and Energy Flow

Below the forest floor lies one of the most sophisticated alignment exchanges in nature. Mycorrhizal fungi don't just attach to roots—they merge ripple fields. The fungus accesses water and minerals from the soil with its vast thread-like network. In return, the plant supplies sugars via photosynthesis. But this isn't barter in the economic sense—it's a harmonic field arrangement. Their root-ripple interfaces synchronize to minimize resource-search tension, making both partners more stable, more efficient. What looks like cooperation is, in fact, mutual alignment maintenance.

Bees and Flowers: Color and Scent as Ripple Resonance Cues

Bees don't "see" flowers like we do—they perceive ultraviolet patterns, magnetic trails, and subtle scent oscillations that act as ripple resonance signals. Flowers, in turn, have evolved precise visual and chemical outputs that match the bee's tuning range. This is no coincidence—it is field-based matchmaking. When a bee lands, the interaction isn't just about pollen—it's about ripple touchpoints between two species designed to complete a resolution circuit. Nectar is just the visible part—the real exchange is vibrational compatibility.

Ecosystems = Networks of Field-Sensitive Organisms Optimizing Tension Equilibrium Together

Zoom out, and you'll see that an ecosystem—whether forest, coral reef, or savannah—is not a collection of organisms. It's a ripple optimization matrix. Each creature, plant, fungus, and microbe is a node in the field, responding to local and global pressures, redistributing energy, adjusting roles, and balancing tension collectively. Predators control overpopulation ripple. Scavengers resolve decay ripple. Decomposers dissolve entropy zones. It's not just food chains—it's field interdependence.

In the Universal Energy Dynamics view, symbiosis isn't generosity. It's recognition of shared alignment opportunity. Life finds harmony not through dominance, but through field-aware reciprocity.

"Symbiosis is not support—it's synchronized survival." "To live together is not just to share space, but to co-resolve existence."

6.6 No Language, No Tools—Just Perfect Engineering

When we look at nature's marvels—the hive of a bee, the nest of a weaverbird, the camouflage of a cuttlefish—we're often struck by the question: *"How do they know?"* No manuals. No teachers. No verbal instructions. And yet, they engineer with a mastery that rivals our most advanced labs.

But what if we've misunderstood the source?

Instincts Are Not Magic—They're Embedded Ripple Memory

The word "instinct" often carries a mystical air, as if animals are guided by some unexplainable inner force. In UED, there's no need for magic. These behaviors are field memory embedded in form. A spider weaving its web isn't "thinking"—it is executing stored resolution pathways, fine-tuned by generations of ripple-informed evolution. The design is not in the spider's brain—it's in the ripple configuration of its being, replaying successful alignment sequences.

Think of it this way: just as ripples stabilize into certain geometric forms under tension, living beings stabilize behaviors that reduce tension in their environment. That's what instinct really is: the replay of previously successful field resolutions.

"Design" in Nature = Encoded Equilibrium, Not Divine Plan

Nature's structures often evoke the idea of a master architect—whether God, evolution, or randomness. But design in UED is not a blueprint from above; it's a self-correcting equilibrium script. Every shape, motion, and behavior is a result of repeated ripple alignment, stabilized over time.

- The honeycomb isn't "designed" to be hexagonal; it becomes hexagonal because that's the most stable pressure-dissipating configuration.

- Bird feathers aren't *invented* for flight; they emerge as optimized ripple diffusers.
- Coral polyps don't *decide* to build reefs—they align with mineral ripples, layer by layer, forming tension-neutral structures.

What looks like intelligent design is, in fact, intelligent resolution—no planning, no foresight, just alignment through persistent field response.

Life Doesn't Solve Problems—It Resolves Tensions

Humans think in terms of *problems and solutions*, often missing the deeper truth: life never sets out to "solve." It simply responds. When there is too much heat, too much hunger, too much imbalance—living systems move toward less tension. That's it. Evolution doesn't favor the fittest in a linear race; it favors those who resolve ripple tension most efficiently.

- A leaf tilts not to "seek" light, but to resolve photon alignment.
- A bird doesn't fly to "go somewhere"—it flies by participating in flow with the least energetic disturbance.
- Even pain is not a signal to fix a problem—it's ripple stress needing alignment.

In nature, there is no tool—but everything is engineering. There is no language—but everything communicates. There is no plan—but every pattern achieves balance.

- "Life does not invent—it aligns."
- "Intelligence does not require language. It only requires awareness of tension."
- "The universe doesn't build engineers. It becomes them through ripple memory."

6.7 What Humans Call 'Instinct' Is Pre-Encoded Resolution Logic

Instinct has long been misunderstood—as something mystical, automatic, or even unconscious. But what if instinct isn't a fallback for low intelligence, but rather a highly efficient shortcut for ripple resolution?

In the Universal Energy Dynamics (UED) framework, what we label as "instinct" in animals—or even reflexes and habits in ourselves—is simply this: pre-encoded alignment patterns designed to resolve tension as fast and efficiently as possible, without the need for conscious deliberation.

Behavior in Animals = Local Field Responsiveness

When a bird lifts off in perfect reaction to wind change, or a gecko drops its tail to escape, these aren't calculated decisions—they're live field responses. Each behavior is born from alignment with subtle ripple conditions in the environment.

- A rabbit doesn't need to "think" about running from a shadow—it resolves tension triggered by ripple distortion in its sensory field.
- A turtle hatching and heading toward moonlight isn't obeying "instinct" like a command—it's following the strongest alignment cue in its local ripple gradient.

This is nature's intelligence: not awareness in the human sense, but pure coherence with field conditions.

"Reflexes" and "Habits" = Internal Resolution Loops

Zoom in on human behavior. Reflexes—pulling your hand from a flame, blinking at a sudden flash—aren't thought about. They're field-protected pathways, hardwired for speed. These aren't decisions. They're ripple shortcuts, paths of least resistance to resolution.

Even habits operate on similar principles. Repeating an action embeds ripple familiarity into your neural and muscular architecture. Eventually, the action becomes the default resolution strategy for certain inputs—whether it's brushing your teeth or reacting with anger.

UED reframes habit not as discipline or laziness, but as looped resolution patterns in the mind-body field. That's why some habits are hard to break: you're not breaking a behavior—you're overriding a resolution loop.

Nature Isn't Conscious Like Us, But It Is Always Field-Aware

Let's be clear: animals, plants, even bacteria don't possess consciousness in the human-reflective sense. But they are always aware—not with thought, but with field sensitivity. That's a more universal, primal awareness. And in some ways, it's more efficient.

- The Venus flytrap doesn't plan to close—it detects ripple interruption and responds.
- A salmon doesn't remember its way home—it follows ripple trail imprints in electromagnetic gradients.
- Cells in your immune system don't "know" you're sick—they respond to ripple disruptions in molecular alignments.

So what humans call "instinct" is not lack of awareness—it's a different form of intelligence, based on alignment rather than abstraction.

- "Instinct is not magic. It's ripple logic, preloaded."
- "Reflexes are not shortcuts of the brain. They are shortcuts of the field."
- "Conscious or not, life is always listening—for imbalance—and answering through design."

6.8 Closing Reflection – Biology Is Ripple Mechanics in Motion

When you strip away the layers of biology—genes, proteins, cells, systems—what remains is something far more elegant: ripple mechanics in motion. Life isn't an accident of molecules, nor a linear script written by DNA. It is an ongoing choreography of alignment, performed by structures that feel the field and resolve accordingly.

The bee building hexagons... The bird surfing wind currents... The root tip finding the gap between rocks...

None of them "know" in the human sense. Yet all of them respond to tension, restructure their form, and navigate ripple gradients with a mastery that engineers still struggle to replicate.

"Life doesn't push through the world. It bends with it."

Unlike machines, life is not forceful. It is sensitive. It does not impose—it adapts, shifts, listens. The ant does not bulldoze its way to food. It follows pheromone trails, which are themselves ripple cues encoded in scent. The banyan tree does not grow in a straight line. It responds to sunlight, shadow, humidity, and space—real-time ripple calculations that no algorithm can match in elegance.

This is not passive surrender. It's active surrender—the kind that aligns, leverages, and flows.

"Every living system is a ripple-aware structure navigating tension intelligently."

What makes something *alive*? It's not heartbeat, or movement, or even DNA. It's the ability to resolve tension through interaction with the field.

- A cell knows where to divide.
- A lizard knows when to flee.
- A vine knows which direction holds the light.

This knowing is not cognition—it's tuned response. And this tuning is not random—it is structured, repeatable, elegant. This is ripple intelligence. And biology is its visible form.

"The only difference between engineering and evolution is that one draws, the other grows—but both resolve."

Engineering sketches the solution. Evolution grows the solution. Yet in both cases, the fundamental question is the same:

How can we resolve this imbalance into alignment?

A jet wing is designed to reduce drag. A bird wing evolves to dance with turbulence. Both are solutions. Both are beautiful. But one requires a drafter. The other—just time and field awareness.

So the next time you see a hummingbird hovering... Or a coral reef forming... Or a spiderweb glistening in morning dew...Don't just admire the biology. See the ripple mechanics beneath it.

"Life is not a miracle of matter. It is the perfection of alignment."

Chapter 7: You Are Not Made of Matter

You Are Made of Ripple Intentions

You were never a thing. Not the bones beneath your skin, not the blood that pulses through your veins, not even the brain wrapped in its skull casing. All of that changes, fades, renews. Yet something endures. This chapter invites you to turn your gaze inward—not to find atoms or particles, but patterns. You are not built from matter, but from ripple intentions—dynamic alignments within a conscious field that sustains your sense of "I." Identity is not substance; it is stability. And the self you think you are? It's not trapped in flesh. It's a resonance, perfectly remembered through tension, coherence, and the quiet will to continue existing.

7.1 You've Never Been a Thing

You've been told you're a body—a machine of cells and circuits. A brain wired for cognition. A structure of carbon, water, and a dash of stardust. But let's pause. Is that all?

Every seven years, nearly every cell in your body is replaced. Your skin regenerates. Your bones remodel. Your gut lining renews. Even your neurons—once believed static—are now known to shift, rewire, and adapt. Yet somehow, *you* persist. Through all this change, the "I" remains anchored. But where?

Here's the shift: What if who you are is not *in* the body, but held *through* it? Not a physical container, but a persistent ripple alignment—a coherent pattern sustained in Structured Vacuum Energy. You are not stored in matter. You are sustained by intention.

This continuity—this felt sense of "me"—is not the result of particles glued together. It is the consequence of an ongoing ripple agreement, a field configuration that remembers how to stay coherent in an ever-shifting universe. You're not an object. You're a resonance. Not static, but stable. Not built, but aligned.

7.2 Identity Is a Ripple Alignment

Who you think you are—your name, habits, memories, preferences, personality traits—is not embedded in bone or flesh. It's a continuously maintained pattern, like the melody of a song played across different instruments, yet somehow always recognizable. The body is not a container of identity—it's the field within which identity stabilizes.

In the Universal Energy Dynamics (UED) framework, every form, including the human body, is a *dynamic ripple field*, not a fixed structure. It is a temporary stabilization of ripple tension within Structured Vacuum Energy (SVE). The bones, blood, and biochemistry are expressions of alignment—not substances carrying identity, but instruments playing it.

Your identity—what you call "me"—is not a single layer, but coherence across many ripple strata:

- **Mind**: The active interpreter, always seeking alignment with intention.
- **Emotion**: The field's feedback mechanism for tension mismatch or harmony.
- **Memory**: A ripple imprint—a record of past resolutions.
- **Attention**: Your field's active alignment vector in the present.

When these layers resonate together, you feel "centered," "yourself," "at peace." When misaligned, you feel scattered, stressed, confused. Thus, identity is not a *thing* but a *state of field integrity*—a condition of resolved tension across your ripple layers.

You are not *inside* your body. You are *stabilized through* it. Like a lighthouse beam that seems to have a location, but is actually the product of alignment across lenses, mirrors, and movement—so too is your identity a consequence of multiple ripple systems syncing in real time. When that alignment holds, you feel whole. When it breaks, you search for yourself.

7.3 Intent = The Invisible Architecture

Intent is often misunderstood as mere *desire*—as something we want, aim for, or chase. But in the framework of Universal Energy Dynamics (UED), *intent* is something far more fundamental. It is the *invisible architecture* behind all form, behavior, and continuity in life.

Intent is not *what* you want. It is *how* the field chooses to organize itself.

Consider this: a cell does not desire to divide, yet it does so with precise timing and structure. A tree does not desire to grow toward the sun, but it aligns its branches in exquisite coordination with light and season. Insects do not strategize their movements, yet they respond in perfect efficiency to changes in the environment.

This isn't randomness—it's ripple intention in action.

In ripple dynamics, *intent* is the inner directional pressure—a directive tension—that causes patterns to form, persist, and evolve. It's what stabilizes the field around a purpose, even if that purpose is not consciously known to the organism. Every act of structure—from protein folding to embryonic development to emotional response—is a ripple resolution guided by intent.

At the human level, intent becomes more refined. Not only do we possess biological intent (like healing, learning, protecting), but we also *steer* that intent through a higher-order field awareness—what we call consciousness.

Consciousness, then, is not the source of intent. It is the *interface* that lets us hold, shape, and direct it.

- When you *focus*, you anchor intent.
- When you *meditate*, you dissolve noise so true intent becomes visible.
- When you act from awareness, rather than impulse, you *align* with field intent—not egoic preference.

This is why you can change your life not by fighting circumstances, but by realigning your inner field of intent. Everything—your behavior, your decisions, your emotional responses—will begin to follow the new architecture you've set.

Intent is the seed of manifestation, but not in the mystical "wishful thinking" way. It is the *structural bias* of the field itself. It doesn't just "wish"—it *shapes*.

So ask not: "What do I want?" Ask: "What is the ripple trying to become through me?" That is the architecture you are here to align with.

7.4 The Illusion of Material Continuity

Every seven years, nearly every cell in your body is replaced. Your skin sheds. Your blood regenerates. Even the bones you believe are static undergo remodeling. So if your "matter" is constantly changing—what remains? What is this *you* that seems so continuous?

This is the great illusion of the material world: that continuity comes from substance. In truth, continuity arises from *pattern*, not particles. From *ripples*, not remnants.

You are not a preserved stack of molecules. You are a repeating ripple agreement.

The field—the Structured Vacuum Energy (SVE)—does not care about pieces. It responds only to *tension* and its resolution. Your body is not a container of stuff; it is a configuration that sustains a certain ripple architecture over time. It's like a song that stays recognizable even if played by different instruments or sung in different keys. The *notes shift*, but the *melody holds*.

That melody—your posture, habits, memories, emotions, even personality—is an emergent coherence of ripple alignments, maintained through intent and feedback. Your cells are simply the temporary scaffolding that hold that ripple shape. And when those cells die, new ones emerge under the same field instructions—unless you consciously alter the field.

This is why you can change. This is also why you resist change.

Change disrupts existing ripple contracts. If your identity is rooted in an old alignment, any shift feels like death—even though *nothing* in your body remains the same anyway. The only thing trying to stay the same... is the *field agreement* about who you are.

That agreement is not bad. It's necessary. But it's also editable.

- You are not "your body." You are its temporary ripple logic.
- You are not "your face." You are the emotional field it responds to.
- You are not "your past." You are the residual alignment echoing through the present.

In UED terms, there is no such thing as a fixed object—only zones of sustained resolution. Your "self" is not material continuity;

it is *intelligent echo*. A pattern held just long enough for the observer to believe, "That's me."

But patterns can evolve. And so can you.

7.5 Feelings, Thoughts, Desires = Field Phenomena

What you call *emotion*, *thinking*, or *desire* is not internal noise—it is field activity. These aren't abstract phenomena occurring inside a "mind-box"; they are real, dynamic ripple events across the body-mind-field interface. They are you—expressed as tension patterns.

Let's break them down:

▪ **Emotions: Ripple Resonance Shifts**

Feelings like fear, joy, sadness, or anger are not floating moods—they are *low-frequency ripple reconfigurations*. When you feel anxious, your field tightens in misalignment. When you're in love, your field expands toward resonance. These are not psychological fogs—they are ripple weather patterns sweeping through your internal field geometry.

Each emotion modifies how ripple tension flows through you:

- Anger = compressed directional pressure
- Sadness = diffused low-gradient stagnation
- Joy = spontaneous field coherence
- Peace = ripple resolution equilibrium

You *feel* emotion because your awareness is tuned to these internal shifts in tension. That's all it is—and it's everything.

▪ **Thoughts: High-Frequency Ripple Activations**

Thinking is not whispering inside your head—it is rapid ripple excitation, mostly occurring in fine-tuned brain–field interactions. A thought arises when certain field alignments spike into temporary coherence. The more aligned the surrounding field, the clearer the thought.

That's why when your field is scattered—tired, stressed, emotionally charged—thinking gets cloudy. Clarity is not intellectual; it's field hygiene.

And no, thoughts don't "come from the brain." They appear in the mind-field, like sparks in an electric fog.

▪ **Desires: Directional Pulls from Unresolved Tension**

Desire is not greed. It is the field's natural impulse to resolve tension. You crave something not because you are broken—but because your ripple field detects a gap in alignment.

- **Hunger** = physical ripple depletion seeking resolution
- **Love** = emotional incompleteness seeking coherence
- **Ambition** = future ripple scenarios pulling your attention for reconfiguration

Desire is not the enemy. It is *directive tension* trying to resolve itself. The problem arises only when the mind misreads the signal and chases substitutions, creating more misalignment.

Core Insight:

You are not thinking, feeling, or desiring *inside* your body. You are a field system undergoing ripple dynamics—always aligning, always responding.

So the next time you feel overwhelmed or ecstatic, remember: This isn't a mood. It's your field shifting shape in real-time. Watch it. Ride it. Align it.

You don't have emotions. You *are* a ripple configuration experiencing resolution.

7.6 The Body as a Conductor, Not a Container

Let's dismantle the oldest illusion: that your body is a closed container, *holding* your life inside like a box of blood and thoughts.

It isn't.

Your body is a conductor—a biological instrument that receives, amplifies, and distributes ripple tension across a larger field. It doesn't *store* your mind or your soul. It *aligns* with them.

Let's reframe each part accordingly:

▪ **The Nervous System: Ripple Amplifier, Not Source**

The brain doesn't generate thought. The spinal cord doesn't *create* sensation.

Your nervous system is like a finely tuned ripple antenna. It takes ambient field inputs—both internal and external—and amplifies their resolution into recognizable experience. It filters alignment feedback from your environment and projects it as sensation, movement, awareness.

Just as a radio doesn't produce music but decodes frequency patterns, your nervous system interprets ripple instructions into perceptual outputs.

▪ **Gut, Heart, Skin – Field-Sensitive Organs**

Modern science is catching up to ancient wisdom:

- The gut is your "second brain."
- The heart has a nervous system of its own.
- The skin can "feel" emotion beyond just texture.

These aren't just poetic metaphors—they are ripple interfaces.

- Your gut is a resolution hub, balancing chemical and emotional signals in low-frequency fields.
- Your heart synchronizes internal rhythm with

environmental ripple coherence.
- Your skin is your outermost field membrane—where ripple exchange meets the world.

Each organ isn't just doing biochemical work—it's modulating ripple tension across your alignment matrix. You literally *feel with your whole body*.

▪ Life Is Not in the Brain—It's in the Balance

You are not "thinking" in the skull and "living" in the chest. You are a distributed ripple system, harmonizing continuously across all zones.

Every cell, organ, and tissue participates in a live feedback loop of tension and release. Life is not a *thing* located somewhere. Life is an ongoing ripple event—always tuning, always conducting alignment across a field.

Core Insight: Your body is not where life *resides*— It's where life resolves.

You are not locked inside a skin-bag. You are a living waveform, *conducting coherence* through matter.

So stop treating the body like a prison. Start using it like the resonant instrument it was designed to be. Feel more. Align better. Because *you don't live in the body*— The body lives through you.

7.7 Ripple Decay = Aging, Misalignment = Disease

Let's take off the lab coat and look at life through the ripple lens—where illness is not an invasion and aging is not a countdown. Both are simply field dynamics playing out.

▪ Coherence Is Health. Its Weakening Is Aging.

Every living system stays alive not by being *built once* and preserved like a statue, but by maintaining alignment in real time.

Your body is a ripple orchestra—countless systems playing in harmony. As long as coherence holds, the melody of life continues: clear, strong, and adaptable.

But over time, as ripple tension attenuates, the resolution becomes less crisp:

- Signals take longer to propagate.
- Alignment becomes harder to sustain.
- The ripple loops that once self-correct begin to lose fidelity.

This is not "decay" in a toxic sense. This is ripple fatigue—a gradual loss of tuning capacity.

Aging, then, is the natural attenuation of alignment efficiency—Not damage. Not deterioration. Just the field whispering more than it sings.

▪ Illness = Misalignment, Not Malfunction

Disease doesn't start in the flesh. It starts where field coherence breaks down.

Before a tumor forms, before inflammation spreads, before organs fail— there is a tension fault in the ripple configuration of that zone.

Misalignment can come from:

- Emotional trauma (disrupting body-field memory)
- Environmental incoherence (toxins, radiation, artificial fields)
- Chronic stress (resonance loops trapped in high tension)

The symptoms are physical. But the cause is always ripple misalignment.

What we call "illness" is the body's field struggling to self-resolve under misaligned conditions.

▪ Healing = Re-Alignment, Not Just Repair

True healing isn't just patching tissue. It's restoring the ripple loop.

That's why:

- Some wounds heal faster with love and rest than with surgery.
- Meditation and breathwork recalibrate the system.
- A peaceful mind accelerates recovery.

The body doesn't fix itself through mechanics alone. It remembers its original alignment and seeks to return.

Even medicine works best when the ripple field is ready to receive it.

Chemicals assist. But healing is coherence re-established.

Core Insight:

"Health is not the absence of pain. It is the presence of resolution." "The body doesn't age—it slowly forgets its perfect rhythm." "Illness is not a fight. It is a misheard song."

You don't need to fear aging or disease. You need to listen—and help the body retune.

Because ripple never dies. It only loses coherence—until you bring it back into balance.

7.8 You Are Held Together by Meaning

You've been taught to search for meaning—as if it were a treasure buried somewhere in the world, waiting to be discovered. But what if that's backward?

What if meaning isn't something *found*—but something that's already *holding you together*?

▪ Every "You" Is a Configuration

At the most fundamental level, there is no *fixed self*. There is no static "me" that was born and remains unchanged.

Instead, there is a configuration—a ripple alignment sustained across time, across memories, thoughts, relationships, sensations.

This configuration is not random. It's coherent, purposeful, and remarkably stable. And the only reason it doesn't dissolve into chaos is because...

You keep intending to exist.

Yes, you—without realizing—are continuously reaffirming yourself:

- Through every choice you make
- Every memory you preserve
- Every morning you wake up and remember "who you are"

That persistent self-alignment isn't mechanical. It's meaning-driven.

▪ The Soul Is Not a Thing

Let's be clear: The soul is not a glowing orb. It's not a weightless traveler between bodies. It's not even a "thing."

The soul, in UED terms, is a persistent ripple alignment—a *longwave coherence* that maintains continuity across moments, lifetimes, dimensions.

It's a thread of intentionality woven through every field interaction that makes "you" recognizable—across all your forms and forgettings.

When people say "you have a soul," what they really mean is:

- "You have a sustained agreement with existence."
- "You are choosing to be here—and that choice shapes reality."

▪ Meaning Is What Binds the Self

We often ask: *What is the meaning of life?* But the better question might be: *What holds life in meaning?*

Meaning is not a conclusion. It is not something to be assigned. It is what enables coherence in the first place.

A meaningless pattern cannot sustain itself. A ripple with no intent collapses into noise.

But a meaningful pattern self-stabilizes. It pulls energy into alignment. It adapts. It remembers. It evolves.

That's you.

Your body, your mind, your values, your presence—are all sustained by the meaning you participate in, even when you're not fully aware of it.

Closing Insight:

"You don't live to find meaning. Meaning is how you continue to live."

"The soul is not searching—it is sustaining."

"You are not held by matter. You are held by meaning."

You are a ripple that *knows why it's rippling*. And as long as you keep remembering that, you'll never fall apart.

7.9 Closing Reflection – You Are a Ripple That Remembers

Pause for a moment—not to observe yourself as a body, but to feel yourself as a pattern.

You've never been the things you thought you were. Not flesh, not bone, not even the thoughts in your head. All of these have changed. All of these have passed.

And yet... *you're still here.*

- **You Are Not Material. You Are Momentum.**

What appears as "you" is not a structure, but a sustained motion of intent—a ripple stabilized across time, holding form without being form.

You are not the particles, but their persistent conversation. Not the notes, but the melody. Not the moment, but the memory of motion through tension.

Matter ends. Motion fades. But momentum carries memory forward.

"You are not material. You are momentum."

▪ **You Are Not Built. You Are Balanced.**

Look at your life—not as a pile of achievements or a fixed personality, but as a balancing act between all the fields you interact with:

- Desires and duties
- Emotions and logic
- Body and awareness
- Self and others

You've never been constructed like a machine. You've been balancing like a ripple—responding, adjusting, aligning.

You aren't a product. You are a process. And that process is *alive*.

"You are not built. You are balanced."

▪ **You Are Not a Collection. You Are a Coherence.**

You've gathered experiences, knowledge, scars, and stories. But they don't define you individually.

What defines you is the coherence across them—the unbroken thread of alignment that makes them meaningful.

You're not the sum of parts. You are the pattern that holds them together.

And that pattern is intelligent. It knows where to flow, what to keep, and when to let go.

"You are not a collection. You are a coherence."

▪ **When Your Alignment Fades...**

Nothing lasts forever in this field of ripple play. There will come a moment when your alignment—this beautiful, delicate coherence—will begin to loosen.

And it won't be a tragedy. It will be a return. A return to the SVE, to the larger field, to stillness. But even then... Your ripple does not vanish. It becomes part of every other ripple it touched. It becomes a memory in the field.

Not a memory in a brain. Not a legacy in a book. But a resonance—still echoing in the balance it once held.

Final Words

"You are not a wave. You are the intention that shaped it. You are not a body. You are the balance it carried. You are not a name. You are a ripple that remembers—even when the world forgets."

Chapter 8: The Difference Between Life and Stone

Life Responds, Stone Follows
LOC for Both—Navigation vs. Submission

At first glance, the difference between a human and a stone seems obvious—one moves, breathes, thinks; the other simply sits. But under the lens of Universal Energy Dynamics (UED), the contrast is far more nuanced—and far more revealing. Both the body and the stone are shaped by the same underlying laws. Both are ripple configurations within the Structured Vacuum. And yet, only one has the capacity to respond, to course-correct, to maintain alignment through intentional tension resolution. This chapter doesn't seek to romanticize life or reduce matter—it redefines both. Life, here, is not an event. It is not a set of chemical traits. Life is the active participation in alignment, the navigation of consequence rather than submission to it. Stone does not lack energy. It simply lacks the will to respond. And that is where life begins.

8.1 What Really Makes Something Alive?

Ask a child, a scientist, or a philosopher: "What is life?" You'll get answers that range from poetic to procedural. Biologists may offer a checklist—cells, metabolism, reproduction, homeostasis. It sounds precise. But it's not satisfying. Because none of these explain *why* something feels alive.

A virus sits awkwardly on that checklist. So does a seed in dormancy. Or a robot that learns. Are they alive? Are they not? The lines blur fast.

What if the problem is not with the examples—but with the checklist itself?

Under the lens of **Universal Energy Dynamics (UED)**, life is not defined by *what it has*—but by *what it does with misalignment.* Life, at its core, is the capacity to respond to the tension of existence.

A stone is perfectly aligned to its surrounding field—but passively so. It accepts consequence. It follows the ripple laws without negotiation.

But life doesn't just follow—it *navigates.* It senses imbalance, resists decay, adapts, evolves. It doesn't sit inside reality—it interacts with it, *intelligently.*

So what makes something alive?

Not just cells. Not just carbon. Not just chemistry. Life is the willingness to participate in resolution. It's the difference between *holding alignment* and *seeking it.*

In other words: A stone submits. Life adjusts. And that shift—from passive alignment to active engagement—is where everything begins.

8.2 The Stone – Perfect Submission to the Field

A stone does not fight. It does not plan, adjust, or hesitate. It simply *is.* And in that state, it reveals something quietly profound.

The stone is not lifeless because it lacks spirit. It is lifeless because it does not resist consequence. It doesn't respond to the world—it *absorbs* it.

When placed under heat, it expands. When dropped, it falls. When weathered, it erodes. It never modifies its path. It never redirects tension. It simply surrenders to every ripple of reality acting upon it.

In Universal Energy Dynamics (UED), this is what we call passive resolution. The stone does not generate new ripple patterns—it is a receiver of alignment instructions from its environment. Its structure reflects a balance—not because it seeks it, but because it has nowhere else to go.

It is stable, yes. But not intelligent. It holds no memory of intention, only the fossil of previous tensions that shaped it.

The stone is not in conflict with its world. But neither is it in dialogue. It is *resolved* only because it offers no resistance—no navigation—no adjustment.

This is why we say:

"Stone = alignment by external resolution."

It is the endpoint of ripple submission. The boundary condition of consequence. Not broken, not flawed—just complete in its surrender.

And yet... This passive perfection is what makes it non-living. Because to be alive is to hold a *choice* within the field.

The stone holds no such choice. It simply follows the ripple path... exactly as given.

8.3 Life – Self-Correcting Ripple Navigation

Unlike the stone, life does not just sit inside the ripple field—it plays with it.

Living systems are not fixed. They are *feedback loops*. They sense changes, detect imbalance, and make constant micro-adjustments to stay coherent. Whether it's a single-cell organism reorienting toward nutrients, or a human making a complex decision under stress—life navigates ripple tension, it doesn't just endure it.

This is the difference: A stone falls. A cat lands on its feet.

Life, at every level, is a self-correcting ripple navigator. It doesn't wait for the environment to dictate terms—it participates in the negotiation. It modifies its own alignment in real time to preserve coherence.

From the bacterium that swims away from acidity to the plant that bends toward sunlight, to the human who rethinks a career to restore inner peace—it's the same principle:

Life = localized intelligence operating through tension feedback.

This is not "conscious thought" as we define it—but it is field awareness. A built-in sensitivity to misalignment and a built-in protocol to resolve it.

In UED terms, this is what we call internal ripple agency. It's not magic. It's not even willpower. It's the innate directive capacity of a system to re-balance itself from within, rather than be shaped entirely from without.

- In non-living systems: tension is resolved *onto* them.
- In living systems: tension is resolved *through* them.

Every heartbeat, breath, blink, movement, decision—whether instinctual or deliberate—is a rippling correction that sustains the illusion of a continuous self. It's not perfection that keeps you alive. It's constant *course correction.*

So the difference isn't in the ingredients. It's in the capacity to respond, recalibrate, and re-align.

That's why:

Life is not the absence of misalignment. It's the art of resolving it on the fly.

8.4 The Law of Consequences (LOC) – Two Modes

The universe doesn't play favorites. Whether you're a rock or a lion, a drop of water or a child dreaming under the stars—you are governed by the same foundational rule:

The Law of Consequences (LOC): All ripple tension seeks resolution, and every action participates in this dynamic.

But how you participate makes all the difference.

▶ **Mode 1: Passive Obedience (Stone)**

A stone has no agency. It doesn't choose. It doesn't interpret. It doesn't respond. It submits.

If a force acts on it, it moves. If it burns, it cracks. If it erodes, it reshapes. In all these cases, the consequences happen *to* the stone—not *through* it.

It is a perfect example of externalized LOC: Resolution occurs by rearranging the environment *around* it until the stone falls into equilibrium.

This is Mode 1:

Passive Obedience. The stone doesn't resist ripple flow. It doesn't contribute to the pattern. It yields.

▶ **Mode 2: Ripple Response Intelligence (Life)**

Now contrast that with life.

Life doesn't merely yield. Life negotiates.

A tree doesn't just get pushed by the wind—it grows into it. A human doesn't just feel hunger—they invent agriculture. A bird doesn't wait for the air to carry it—it rides thermals, adjusting wing tension on the fly.

This is not defiance of nature—it's intelligent cooperation with ripple dynamics. Life is the active partner in LOC.

Consequences don't just happen *to* it—life influences the path of those consequences.

This is Mode 2:

Ripple Response Intelligence. Where the living system becomes a field-aware participant, shaping its fate through constant realignment.

So while both stone and life exist within the same ripple law, their roles differ dramatically:

- Stone obeys the ripple.
- Life responds to it.

One submits to consequence. The other navigates it.

And that is what makes life... *life*. Not chemistry. Not consciousness. But the capacity to respond with alignment in the face of ripple disturbance.

"To live is not to avoid consequence. It is to flow with it, feel it, and forge meaning from the resolution."

8.5 Resistance vs. Response

The universe speaks through ripple tensions. Everything—living or not—feels the press. But not everything knows what to *do* with it.

And that is where the divergence begins: Resistance versus Response.

▸ **The Stone: Resistance Until Failure**

When a stone is struck, it doesn't negotiate. It doesn't adjust, compensate, or redirect force. It resists—blindly, uniformly, stubbornly.

Up to a point, this works. The lattice of its atomic alignment *holds*. But as pressure builds, internal stress accumulates, and eventually—

The stone breaks.

This is the limit of passive obedience: It holds... until it can't.

There's no intelligence in its rigidity—only the temporary illusion of strength. The stronger the resistance, the closer it is to fracture.

▶ **Life: Response Until Adaptation**

Now observe life.

A sapling in a storm doesn't resist the wind—it sways. A muscle under stress doesn't snap—it strengthens through micro-tears. An organism facing heat doesn't fight it—it sweats, disperses, and regulates.

Life is not passive. But it is not rigid, either.

Life doesn't oppose the field—it enters into dialogue with it. Tension is not a threat—it's feedback.

Life doesn't resist to survive. It responds to evolve.

This is ripple intelligence in action: Micro-adjustments, recalibrations, feedback loops—minute by minute, heartbeat by heartbeat.

▶ **Bending, Not Breaking**

The deeper truth is simple:

- The stone appears strong but is inherently fragile.
- Life appears soft but is resilient beyond measure.

Because resistance relies on holding against the ripple. But response flows *with* it—without losing identity.

"The stone breaks because it refuses. The tree survives because it listens."

▶ **Response ≠ Emotion**

A key clarification here: Response is not the same as reaction, or emotional fluctuation.

- A reaction is instinctive.
- An emotion is an interpreted wave.
- But a response is deliberate ripple recalibration—a choice

to adapt the alignment.

It's the difference between collapsing under pressure...And reorganizing your pattern to handle it.

Final Insight:

"The stone resists and falls. Life responds and rises."

Where resistance ends, life begins. And in every breath, every heartbeat, every blink—you are not holding back the ripple...You are *dancing with it.*

8.6 The Soul Threshold – Where Ripple Gains Intent

There is a moment in the ripple spectrum where something changes. Where alignment doesn't just happen—it wants to happen again. Where balance becomes not just a state, but a goal. That moment marks the emergence of what we call life.

That moment is the Soul Threshold.

▸ Not a Metaphysical Add-On—A Physical Intent

Let us first clear the table. The soul is not a glowing orb. It's not something added to a body. It's not an ethereal guest from another world.

In the UED framework, the soul is ripple behavior—when structured vacuum tension becomes self-preserving. It is a field coherence that begins to resist entropy—not with brute force, but with intentional re-alignment.

Life begins not when cells appear, but when a ripple configuration begins to *sustain* itself against dissolution.

▸ Stone Never Cares

A stone erodes. Breaks. Disperses into dust. Not once does it try to hold itself together.

It follows the Law of Consequences (LOC) passively. If the environment erases it, so be it. There is no protest. No pause. No preservation.

A stone doesn't care if it continues to exist. Because it can't care—it has no ripple intent.

▸ **But Even Bacteria Do**

Now look at the smallest form of life. A bacterium doesn't know your name. It doesn't believe in God. But it will do everything it can to avoid death. It will resist antibiotics. Adapt. Divide. Multiply. Survive.

Why?

Because something inside its field says: *Hold the pattern.*

This is not thought. This is not consciousness in the way humans imagine it. This is proto-intelligence—a primitive but unmistakable intent to exist.

▸ **The Soul Is Alignment That Fights to Stay Aligned**

So what is a soul?

Not a substance. Not a metaphysical construct. But a ripple configuration so stable, so coherent, it remembers itself, and it wants to continue.

"The soul is not an insert. It is an emergence."

Where the stone merely submits to consequences, Life begins to navigate them. That navigation, sustained over time, becomes the soul.

▸ **Final Thought: A Threshold Crossed**

The stone and the soul are not opposites. They are points on a continuum of ripple intent.

- The stone: aligned but passive.
- The plant: aligned with responsive structure.
- The animal: aligned with directed sensing.
- The human: aligned with meaning, memory, and choice.

Somewhere in between—the threshold is crossed. And the ripple begins to care if it lives.

That's when matter becomes meaningful. And reality starts to feel.

8.7 Death – The Collapse Back Into Stone-Like Submission

Death is not the opposite of life. It is the end of ripple participation. Where once a configuration held itself together through subtle tension intelligence, now... it lets go.

▸ The Disbandment of Alignment

At the heart of life is an effort— A quiet but constant recalibration to maintain structure, purpose, and presence. That effort is intent. Not conscious in the way humans define it, but directional field behavior that says: *stay coherent*.

When death occurs, this intent ceases. Not in a sudden explosion, but in a withdrawal of alignment. The body doesn't vanish. The particles don't evaporate. But the *ripple architecture* that held everything in meaningful shape... collapses.

What remains is not a person. Not a life. What remains is what was always secondary: Structure without purpose. Form without navigation.

▸ The Body Becomes Like the Stone

Just as a stone doesn't choose to exist, The dead body no longer holds any self-preserving will. It becomes a passive node in the field, subject to decomposition, temperature, decay, and external forces.

There is no more feedback loop. No more tension correction. No more ripple resistance.

Like the rock, it follows consequences without protest.

"Death is not when the body stops working. Death is when the ripple gives up responsibility."

▸ No Disappearance—Only Dispersal

In Universal Energy Dynamics (UED), death does not equate to disappearance. Nothing is truly lost. What ends is the coherence—the specific configuration of ripple tension that *used* to mean *you*.

The components remain. But the alignment—the intelligence that orchestrated them—dissolves.

This is why the body looks eerily similar in death. It's not disintegrated. It's simply abandoned. No longer directed. No longer tuned. It begins to submit, just like the stone.

▸ The Silence of Intent

Perhaps the most haunting difference between life and death isn't visible. It's not the coldness. Not the stillness. It's the absence of *intention*. That animating field pattern that once held the form in negotiation with entropy—is gone.

"Death is not the end of matter. It is the end of matter that mattered."

▸ In Summary: The Great Ripple Fall

- Life = field response, self-sustaining intention
- Death = field submission, absence of navigation
- Stone and corpse = same ripple state: passive alignment without participation

Death is not a mystery in UED. It is the collapse back into field obedience. From coherence to disbandment. From ripple agency... back to ripple silence.

8.8 Why Life Feels Meaningful, and Stone Does Not

Meaning is not found in the shape of a thing. It is found in its ability to respond, to feel tension, and to seek resolution. This is the silent miracle of life: it doesn't just exist—it participates.

► **Life Experiences Alignment, Tension, Fulfillment, and Choice**

A living system is not just a configuration of atoms.

It is a ripple-aware entity, constantly negotiating with its environment.

- It feels alignment when it is safe or stable.
- It senses tension—internal or external—as a signal to adjust.
- It seeks fulfillment—not as desire, but as ripple equilibrium.
- And it makes choices—each movement, breath, reaction is a decision, a micro-expression of autonomy within the universal field.

This play between tension and resolution is what gives life its depth. It is the ripple game of responsiveness. And it is this game that gives rise to the experience of meaning.

► **Stone Is Silent—Not Lesser, Just Uninvolved**

A stone isn't meaningless because it's inferior. It's meaningless because it is not involved.

It does not seek. It does not respond. It simply sits in passive compliance with its surroundings.

It is aligned—but not *intentionally*. It holds form—but not through effort. It obeys the Law of Consequences—but without agency.

"Meaning arises not from existing, but from participating in existence."

▶ **Living Is the Privilege of Navigating Existence with Awareness**

What makes life precious is not its ability to move or grow. It is its ability to sense itself inside the ripple field and to act upon that sensing.

A tree does not speak, but it reaches. A bacterium does not think, but it redirects. An animal does not write, but it reacts. A human does not just act—it reflects on action.

Each layer of life is an increasingly refined ripple interface with the universe.

This navigation—however primitive or profound—is what makes life *feel alive*.

▶ **This Is Why We Value Life—Because It Participates**

Why do we mourn a dead body but not a broken rock? Why do we protect an injured bird but not a falling leaf?

Because intuitively, we sense the difference: Life is not defined by form—it is defined by response. It is not the presence of matter, but the presence of mattering.

A stone doesn't try to continue. Life does.

A stone doesn't care about outcome. Life does—even when it can't explain why.

This is why life stirs emotion, evokes compassion, and demands ethics. Not because it is more beautiful... but because it is more involved.

"Meaning is not a gift. It is a consequence. Only that which participates in tension has the privilege of fulfillment."

Final Thought:

A stone fulfills a shape. But only life fulfills a *purpose*. Because only life walks the razor's edge— between alignment and collapse, between stillness and becoming. Between ripple... and intent.

8.9 Closing Reflection – Life Is Not a Privilege, It's a Promise

Life is not something *granted* to you. It is something held, moment by moment—by your choice to respond.

A stone exists without needing to try. It obeys ripple laws passively, absorbing and releasing tension like a forgotten drum in the wind. But life? Life plays the drum. It listens, adjusts, echoes—and sometimes, it dances.

▸ **"You Are Not Alive Because You Breathe. You Are Alive Because You Remember to Respond."**

Breathing is mechanical. Even a machine can mimic it. But response—real response—is awareness meeting tension.

You are alive not because blood flows. You are alive because you feel the shift, and you move with it.

You respond to pain. To joy. To change. You participate in a ripple field that never pauses, and your very being is a series of micro-decisions: to align, to adapt, to express, to remember.

That memory of response—your intent to continue participating—is life.

▸ **"The Stone Obeys. You Choose."**

The stone has no say in its path. Gravity pulls it. Temperature erodes it. Time wears it.

It follows the field in perfect submission. It never resists. But it also never *becomes*.

You, on the other hand, can shift your path. You can resist decay, redirect forces, realign pain into purpose.

You can say no to the consequence and yes to new alignment. You can refuse entropy with an act of will. You can engineer coherence in the middle of chaos.

That power is not just a privilege. It is the promise embedded in ripple intelligence.

▸ "The Same Ripple Laws Govern Both. But Only One Turns Them into a Dance."

Stone and self—same physics. Same SVE. Same ripple laws. Same universal tensions.

But the outcome? Radically different.

Because while the stone collapses into form, life emerges into meaning.

Where the stone endures, you express. Where the stone falls, you fly. Where the stone is dragged, you navigate.

This dance—this choice to move with the music of the universe rather than just echo it—is what defines a living system.

The ripple laws don't favor the living. They simply offer a stage. And you—by choosing to respond—become the choreography.

Final Note:

Life is not something you possess. It is something you perform. And each moment you remember this—not just to exist, but to engage—you fulfill not a gift, but a cosmic agreement:

That ripple will rise. That tension will come. And that you—unlike the stone—will rise to meet it.

Not because you must. But because you can.

Interlude: The Intelligence Beneath Biology

"Every cell knows what you forgot."

You are not a miracle. You are an *arrangement*.

Part I dismantled the illusion of external force and revealed a universe shaped by structured stillness—Structured Vacuum Energy (SVE)—where everything emerges from ripple equilibrium, not mechanical causality.

Now, in Part II, the ripple turns inward.

This is where science ends and life begins. Not because the rules change, but because something within them remembers.

Life is not the opposite of death. It's the opposite of indifference.

Every ant, bee, bird, and banyan tree is engineering—not by design, but by ripple intent. They don't need equations. They don't need to *understand* balance. They are balance.

They feel alignment the way we feel emotion. They respond to tension the way we respond to music. Not by thinking. By being in tune.

In these next chapters, you'll confront a new truth:

You are not made of matter. You are made of memory—the memory of coherence, of ongoing participation in a ripple field that never sleeps.

Atoms are not building blocks. They are field agreements. Cells are not machines. They are alignment artists. You are not flesh and bone. You are intent suspended across time.

And when you compare yourself to a stone, you'll see it's not substance that defines you.

It's whether you respond.

Stone follows the ripple. Life feels it. You—when awake—can shape it.

This part of the journey will not just redefine biology. It will dissolve the boundary between life and physics, and between *you* and the universe you thought you lived in.

The field never forgot you. Only you forgot how to listen.

Let's listen again.

Part III: The Law of Consequence (LOC)

The final authority in a ripple universe.

Chapter 9: Karma Is Physics

LOC as the True Karma Law — Ripple Feedback, Not Reward/ Punishment

What if everything you've ever believed about karma was a metaphor for something far more precise? Strip away the morality. Discard the myths of cosmic judgment or divine bookkeeping. What remains is a law—neutral, flawless, ever-present. Karma is not a moral force. It is the physics of ripple consequence. In a universe governed by alignment and tension, the Law of Consequence (LOC) operates not through punishment or reward, but through perfect ripple feedback. Every thought, action, or intent introduces a shift in the field—and the response is mechanical, not mystical. This chapter reframes karma not as fate, but as field behavior: an echo of your own misalignments, a mirror of your ripple interactions. You are not being judged—you are being shown.

9.1 What If Karma Was Never Moral?

Karma is often misunderstood as a moral weighing scale—a cosmic system of rewards and punishments that keeps track of your virtues and vices. This popular belief, however poetic, distracts from something more immediate, more scientific, and more exact: karma is not judgment—it is response.

In Universal Energy Dynamics (UED), we discard the lens of morality and look instead through the clarity of ripple mechanics.

Every action, thought, or intent is a disturbance in the structured vacuum—a ripple that moves outward, meets resistance, reflects, refracts, or gets absorbed. The result? A field consequence, not a divine reaction.

Karma is not about good or bad. It's about alignment or misalignment. You are not being punished by some external lawgiver. You are simply receiving the echo of what you set in motion. And because the universe is precise—not sentimental—the echo is not a verdict. It is a feedback loop. The ripple you generate is the ripple you navigate.

The old question—"Why do bad things happen to good people?"—begins to dissolve when we realize that karma is not checking your resume. It is reflecting your current state of alignment. Morality is a human map. Karma is the terrain.

So ask not: "Am I being punished?" Ask instead: "What ripple am I swimming in?"

9.2 LOC – The Real Engine Behind Karma

Behind every event, experience, or outcome in your life is not a god tallying up your deeds. It's not fate playing dice. It is the Law of Consequence (LOC)—the foundational mechanism of the ripple universe.

In the framework of Universal Energy Dynamics (UED), LOC replaces myth with mechanics. It describes how every ripple you generate must resolve, not by choice, but by necessity. The Structured Vacuum Energy (SVE) field doesn't forget, forgive, or favor. It simply responds—perfectly, proportionally, and without bias.

Here's how it works: Every intent, emotion, thought, or physical action sends a tension pattern into the field. That pattern moves outward, interacts with the field environment, and generates a consequence—a ripple feedback that comes back to the source or flows into the surrounding system.

This is not divine punishment. It's ripple resolution. You didn't "do something wrong"—you triggered a field imbalance. And now, LOC is restoring equilibrium. That restoration feels like life: sometimes joyful, sometimes painful, always precise.

So instead of asking "Why is this happening to me?" The real question is: "What ripple did I release—and how is it returning?"

There's no overseer watching from above. The system is self-correcting. Karma is just physics wearing ancient clothes. And once you understand LOC, you begin to see that nothing is personal—everything is patterned.

9.3 Ripple = Action, Tension = Result

Karma is not about what you *meant* to do. It's about what you *did* to the field. In a ripple-based universe, there is no cosmic judge—just consequences calibrated by energy dynamics.

Every action—mental, emotional, or physical—creates a ripple in the Structured Vacuum Energy (SVE). That ripple is not a metaphor; it's a literal field disturbance that shifts local tension. And the universe, by its very architecture, must resolve that tension. This means your ripple triggers a field correction—not because you were wrong, but because equilibrium was disrupted.

You don't "cause" outcomes in the sense of a linear chain of dominoes. You release a ripple. That ripple expands, interacts, resonates, collides, and eventually returns—or resolves through indirect pathways. This is why karma often appears delayed, diffuse, or confusing. It's not a scoreboard; it's a dynamic tension-feedback loop.

Importantly, the field does not respond to your intention—it responds to your ripple's intensity and structure. You may have meant well, but if the ripple carried confusion, resistance, or aggression, the consequence reflects that—not your goodwill.

So, in UED terms:

- Ripple = Action: Every disturbance introduced into the field.
- Tension = Result: The balancing consequence that follows.

It's not about fairness or retribution. The universe isn't punishing you. It's simply balancing you.

9.4 Misalignment Echoes Back—That's Karma

Karma is not a system of judgment—it's a system of field symmetry. What you send into the ripple field—whether constructive or disruptive—returns not because of morality, but because of mechanics.

Every harmful act, whether it's an outburst of violence, manipulation, deception, or even internal resentment, introduces a misaligned ripple into the Structured Vacuum Energy (SVE). This ripple disrupts the coherence of the local field. It doesn't matter who witnessed it or whether it was hidden. The field felt it. And because the field is always seeking resolution, that misalignment must echo back—not as revenge, but as a natural correction.

On the other hand, helpful, caring, or harmonizing actions also ripple out. But these waves reinforce field coherence. They don't just avoid misalignment—they amplify balance, often leading to better health, clearer relationships, and less resistance from the world.

You don't receive consequences because you were "good" or "bad." You experience what you shape in the ripple fabric—just as a sound returns in an echo chamber, changed only by the structure that reflected it.

There's no scoring system. Only return paths. Only field memory. Only you, shaping your experience by how well your actions align with the universal pattern.

This is why the Law of Consequence is more reliable than any moral framework. It doesn't care what you believe. It only cares how you ripple.

9.5 No Reward, No Punishment—Just Field Behavior

Karma is often misunderstood as a cosmic reward system—one that doles out pleasure or pain based on whether your actions were "good" or "bad." But from a Universal Energy Dynamics (UED) perspective, karma is not judgment—it's a ripple equation.

The Law of Consequence (LOC) does not function on ethics, morality, or social approval. It functions on alignment. When your actions generate tension that resonates with the field, you experience ease, flow, and support. When your actions generate dissonance, you experience resistance, friction, and breakdown—not as a moral punishment, but as ripple feedback from a disrupted field.

"Good" and "bad" are simply labels humans assign—emotional wrappers for what is actually just energy behavior.

This also explains why selfish people often succeed—at first. Their actions may be aligned with their immediate field: focused intent, calculated decisions, minimal doubt. The field doesn't judge their morality; it only responds to the coherence or misalignment of their ripple signature.

But over time, when selfishness creates increasing imbalance—betrayed trust, broken systems, internal guilt, emotional suppression—the ripple distortion becomes too great. The delayed misalignment builds beneath the surface until consequence unfolds not from ethics but from structural failure in their personal field.

So yes, results come to those who are sharply aligned with their own intent, even if it's self-serving. But results stay only with those who maintain field harmony across self, others, and context.

Karma doesn't reward you. Karma doesn't punish you. Karma responds to your ripple configuration, not your character description.

It's not what you believe. It's what your energy does.

9.6 Time Delay in Karma = Ripple Lag

One of the most misunderstood aspects of karma is its timing. People ask: "*Why do bad people seem to thrive while good people suffer?*" or "*Why didn't anything happen to them after what they did?*" The answer lies not in divine justice, but in ripple geometry.

Under the Law of Consequence (LOC), karma is not instant because ripple fields don't resolve in straight lines—they propagate through layered, entangled fields. Like dropping a stone into a pond, the wave doesn't bounce back instantly; it moves outward, interacts with other waves, and only then returns as feedback. That delay is not failure. It's the physics of field resolution.

Your action—be it a thought, decision, or behavior—generates a ripple. That ripple must pass through layers:

Your own internal field (emotions, mind, body)

Your social field (relationships, networks)

The environmental or systemic field (community, ecosystem, economy)

The universal field (larger pattern beyond perception)

Each of these layers modulates the ripple, either absorbing, amplifying, or redirecting it. And because these systems have different response speeds and densities, the consequence takes time to arrive. This is why some karmic returns feel immediate, while others surface years—or even lifetimes—later.

You're not being ignored. You're not being spared. You're being scheduled.

Karma is not a lightning strike—it's a returning resonance. And the field never forgets.

So when nothing seems to happen... Don't assume resolution isn't in motion. It is. Just not in your preferred frame rate. Because karma doesn't operate on your clock—it follows ripple geometry, field tension, and alignment thresholds.

In short: *"Karma is not late. It's just working through geometry."*

9.7 Collective Karma – Shared Ripple Contracts

Karma is often seen as a personal affair—a one-to-one relationship between your action and your outcome. But in a ripple-based universe, no ripple exists in isolation. Every ripple overlaps, interferes, amplifies, or cancels another. This makes karma not only personal—but collective.

From families to civilizations, groups share ripple fields. This is not metaphor. It's field mechanics.

- A nation generates a collective ripple by the actions, values, decisions, and alignments of its people and leaders.
- A species generates karma through the long-term treatment of its environment, other species, and its own kind.
- Even organizations, religions, and ideologies carry ripple memory—energetic imprints of how they operated and influenced others.

These ripple fields become shared contracts. You're born into them. You inherit them. Not in the sense of blame, but of influence. Your individual actions are shaped by a pre-existing tension field, just as your DNA is shaped by lineage. But that doesn't mean you're powerless.

You can:

- Perpetuate the misalignment (echo the old patterns)

- Avoid the field altogether (passive withdrawal)
- Or intervene with awareness (become a point of resolution)

History, then, is not just a record—it's a field of unresolved ripple loops. Wars don't die. Injustices don't vanish. They stay unresolved in the collective field until someone—or many—choose to realign.

This is why:

- Trauma repeats through generations.
- Revolutions often mirror the regimes they replace.
- Collective denial leads to systemic collapse.

But also why:

- A single act of deep alignment (truth, compassion, justice) can shift entire fields.
- Movements rise when individuals reclaim field agency.
- Healing is contagious—if initiated with awareness.

In this view, collective karma is not fate. It is field inheritance with participatory terms.

You didn't choose the ripple field you were born into, But you contribute to how it evolves.

So the question is no longer: "What karma am I carrying?" But rather: "What field am I continuing—or correcting?"

9.8 Self-Healing Karma – Corrective Alignment

Karma, through the lens of ripple-field dynamics, is not a punishment—it's a self-correcting feedback loop designed by the universe to restore equilibrium. When misalignment occurs—whether in thought, action, or intent—the field doesn't retaliate. It responds.

Think of it this way: A ripple, once introduced into a stable field, will disturb its balance. The universe, governed by the Law of Consequence (LOC), immediately begins to compensate—not out of vengeance, but through a natural tendency toward resolution.

That's what karma is: A call for recalibration.

Suffering is not the objective. Awareness is.

The discomfort you feel is not the field's way of shaming you—it's its way of informing you:

- That something is misaligned.
- That your output has created discord.
- That there is a better path back to harmony.

Just as pain is the body's ripple alert for injury, karma is the field's ripple alert for energetic deviation.

And just like a wound heals when you stop aggravating it and restore proper flow, karma begins to self-heal the moment you re-align.

This is the most overlooked truth:

Karma is not fixed. It is fluid. It heals.

It doesn't demand retribution; it offers opportunity:

- To revisit the ripple you set in motion.
- To realign your intention with consequence.
- To evolve your expression of self.

This is why awareness alone begins to shift karma.

Because the field is intelligent. It doesn't need repentance. It needs resolution.

So when you notice a repeating loop in your life— A toxic pattern, a recurring conflict, a chronic discomfort— Don't ask, "Why is this happening to me?" Ask instead:

"What ripple am I still misaligning?" "What resolution is the field inviting?"

This is evolutionary karma—not inherited guilt. It's a dynamic invitation to become coherent with reality.

You don't fix karma by fearing it. You fix it by feeling it, facing it, and flowing differently.

9.9 Closing Reflection – Karma Is Not Watching You. It Is You.

For centuries, karma has been imagined as a cosmic judge, watching from above, tallying up your rights and wrongs, preparing consequences like a divine accountant. But this is a myth—one born from fear and maintained by misunderstanding.

Karma is not watching you. It is you.

You are not being evaluated by an external law. You are the very field of ripple interactions that creates, absorbs, and echoes your own consequences.

When you act, think, or intend, you disturb the stillness around you. That disturbance doesn't vanish—it enters the universal memory. It interacts with all other fields—yours and others'—until it finds its resolution.

"You are not under karma. You are the echo it remembers."

This isn't metaphor. It's physics. In Universal Energy Dynamics (UED), all actions—internal or external—introduce tension differentials into the ripple field. These tensions seek balance, and they do so through a path that often looks like consequence.

But there's no punishment involved. No cosmic courtroom. Just the law of energetic memory and correction.

"The universe doesn't punish. It resolves."
This means:

- What you experience is not fate—it is feedback.
- What you carry forward is not debt—it is dissonance.
- What you must do is not repent—but realign.

Karma is not a test you pass or fail. It is not someone else's rulebook you're trying to obey. It is your own ripple loop, finding its way back to stillness.

"Karma isn't a test. It's your own tension—returning home."

So the next time you feel caught in consequence, don't search for blame. Don't fear divine punishment. Instead, pause—and listen to the field.

What is it echoing back to you? What misalignment is it offering to resolve?

Because karma doesn't follow you. It flows through you—a perfect mirror of what you've introduced.

You are not being watched. You are being remembered.

Chapter 10: You Don't Break Nature's Law—You Misalign

Failure as Misalignment. Correction Through Understanding Ripple Tensions.

You were never meant to obey nature's laws—you were meant to align with them. This chapter invites a fundamental shift: stop thinking in terms of breaking rules and start seeing in terms of breaking alignment. In a ripple-based universe governed by the Law of Consequence (LOC), outcomes are not punishments—they are precise echoes of your degree of coherence or distortion. Failure isn't a moral fall. It's a signal. A ripple misfired. A tension unresolved. You don't fall *from* grace—you fall *out* of alignment. And the path forward isn't about atonement. It's about recalibration. The universe isn't watching you—it's responding to you. Always.

10.1 Laws Can't Be Broken, Only Misused

We've long lived with the myth that nature's laws can be broken—like traffic rules or commandments written in stone. But in the ripple universe governed by the Law of Consequence (LOC), this framing is obsolete. Nature's laws aren't rules you obey out of fear. They are structures you align with—or suffer distortion.

No one "breaks" gravity. They misalign with its behavior and experience its result. You don't break thermodynamics. You create inefficiencies that ripple back as loss, heat, or decay. You don't break karma. You introduce misaligned tension into the field—and the field remembers.

The universe doesn't get angry when you misstep. It simply echoes your input through its architecture of balance. Every negative result is a feedback loop, not a punishment. It's nature showing you what doesn't fit.

Misalignment, not rebellion, is the cause of failure. Correction, not remorse, is the path forward.

In this new lens, you are not a violator of laws—you are a tuner of alignment.

10.2 Misalignment = Ripple Dissonance

In Universal Energy Dynamics (UED), *misalignment* isn't sin—it's signal.

At its core, misalignment is a dissonance between your intent and the natural field condition around you. It's not about doing something "wrong" in a moral sense. It's about applying effort, direction, or energy in a way that does not harmonize with the current ripple geometry of the environment, the body, or the situation.

Think of it this way:

- A bird flapping against a storm doesn't "break the law of flight"—it simply fails to align with the wind's rhythm.
- A person pushing through exhaustion to meet a deadline may succeed temporarily—but the misalignment with the body's ripple field will echo later as illness or burnout.
- An argument escalates not because people are evil—but because unconscious speech patterns generate opposing ripple tensions, feeding dissonance rather than resolution.

Misalignment shows up in many forms: wrong timing, forced effort, poor attention, reactive decision-making, or pursuing outcomes disconnected from current field reality. You feel it as stress, fatigue, conflict, confusion, or stagnation.

And here's the key insight: Nature doesn't punish misalignment. It amplifies the unresolved tension until you pay attention. That's what you call failure, frustration, or fate.

The Law of Consequence (LOC) is not retribution—it's resonance. It will keep ringing the same note louder and louder until you *tune yourself* back into alignment.

Misalignment isn't the end of the story. It's the start of awareness.

10.3 What We Call Failure Is a Field Correction

Failure, in its deepest sense, is not a flaw. It is *field feedback*.

In the framework of Universal Energy Dynamics (UED), failure occurs when a ripple—an action, intention, or expression—forces its way through a field condition it isn't aligned with. It might initially appear to move forward, but it lacks sustainability. It cannot maintain coherence. The system begins to resist, distort, or break apart—not because the universe is against you, but because the ripple you introduced cannot integrate with the environment you're in.

That's why *real failure* is not falling down. Real failure is staying out of tune with your surroundings, persisting in dissonance, and refusing to readjust. You may keep succeeding on paper while slowly accumulating inner and outer tensions—fatigue, resentment, isolation, or systemic instability. Eventually, the field corrects you.

Failure is *not the end*. It's nature's way of introducing friction to slow you down before you go too far off-path. It acts like a guidepost—not to shame you, but to realign you. Insects fail, plants fail, even entire ecosystems fail temporarily—but they always respond by reorganizing. That's how evolution works—not as a ladder, but as countless course-corrections.

When you listen to failure as field correction, not personal defeat, you unlock its intelligence. You stop blaming yourself or others, and start reading the feedback. You don't ask, *"What did I do wrong?"* You ask, *"Where did the ripple break coherence—and how can I restore it?"*

Failure isn't a curse. It's a calibration.

10.4 Alignment Is Not Perfection—It's Resolution

In the ripple universe, there is no such thing as perfection. There is only *resolution*.

Perfection implies a final, flawless state—untouched, unchanging, and absolute. But nature does not operate through perfection. It evolves, adapts, adjusts. It doesn't aim to freeze an ideal—it aims to *flow with tension* until coherence is achieved. That's alignment.

Alignment is a moment-to-moment phenomenon. It's the art of listening to tension and responding—not rigidly, but rhythmically. Like a flock of birds weaving mid-air, or a tree leaning with the wind, alignment is about attunement—not dominance. It doesn't seek to overpower reality but to *resolve into it*.

This means mistakes are not failures of character—they're *essential feedback*. They signal where ripples have gone out of sync. They're not events to be hidden or regretted; they're invitations to recalibrate. Life doesn't evolve by getting everything "right." It evolves by trying, sensing, adjusting, and refining.

Think of alignment as musical. A musician doesn't play perfectly every time—but listens, adjusts pitch, balances timing. Similarly, living in alignment isn't about getting it all correct in one go. It's about *tuning yourself continuously*—to your surroundings, your values, your intent.

Perfection is a static illusion. Alignment is a living dance. And in that dance, every misstep teaches rhythm.

10.5 Force vs. Flow – The Misuse of Energy

Misalignment is not always a result of ignorance—it's often a consequence of *misapplied energy*. And the most common mistake? Choosing force over flow.

Force is the attempt to override the field. It's when you push against conditions rather than reading them. It's when your desire dictates the move, not your awareness of the ripple tension. You try to make things happen by will alone, instead of *aligning with the feedback already present*.

Every system in nature avoids unnecessary force. Trees don't force their way through stone—they find the cracks. Rivers don't break mountains in a day—they erode, adjust, and curve. Animals don't exhaust themselves—they conserve, wait, strike with minimal effort. Nature flows. Humans, too often, force.

Force is expensive. It burns energy faster, creates resistance, and leaves behind imbalance. And when the force fails, we call it frustration, burnout, breakdown, or even fate. But really, it's just feedback: *you're misaligned*.

Flow, on the other hand, is the intelligent use of ripple feedback. It's the art of matching effort to field receptivity. It doesn't mean passivity. It means *precision*. The right action, at the right time, with the right pressure—produces outcomes that seem effortless.

This is why alignment is the highest efficiency. It's not about doing less—it's about *wasting nothing*. Aligned actions create minimal resistance and maximal stability. You don't "fight" the world. You *navigate* it, like a surfer riding the wave rather than swimming against it.

In Universal Energy Dynamics, energy isn't a brute force—it's a negotiation between intent and environment. When you apply it with ignorance, it becomes friction. When you apply it with attunement, it becomes flow.

So the choice is constant: Force and exhaust—or flow and evolve.

10.6 How Nature Corrects Without Blame

Nature never says, "You failed." It simply says, "Try another way."

In the ripple universe, correction is constant, but blame is nowhere to be found. A tree growing on a windswept hill doesn't curse the wind—it bends. A river encountering a boulder doesn't blame the rock—it reroutes. A predator missing its prey doesn't sulk—it sharpens its instinct or hunts again.

Nature adjusts. That's all.

This is a key difference between natural intelligence and human interpretation. Nature doesn't assign shame, guilt, or regret. It has no narrative of *good* or *bad*. It operates on the simple, elegant logic of feedback and realignment. Misalignment leads to tension. Tension invites correction. The process continues until equilibrium is found.

But humans—oh, we complicate it. We take misalignment personally. We turn natural feedback into failure stories. We punish ourselves for choices that didn't work, even when they were just unfit for the field at the time. And in doing so, we delay the very resolution we need.

This is where the Law of Consequence (LOC) offers deep clarity. It doesn't punish—it reflects. It doesn't blame—it balances. If the echo of your action returns unpleasantly, it's not to judge you. It's to inform you. It's the ripple's way of saying: *"Your current trajectory needs refinement."*

Healing begins the moment you stop asking *"Whose fault is this?"* and start asking *"Where is the misalignment?"*

When you see yourself as part of a ripple field—not a fixed entity—you stop defending your position and start adjusting your posture. You become like nature: curious, responsive, fluid.

Blame is stagnant. It clogs the ripple with emotional noise. But alignment is always moving. It listens. It learns. It restores.

So the next time something breaks, bends, or bruises in your life, ask not who is to blame—but what needs to shift. Because nature's intelligence isn't about fault. It's about form—reshaped, until it fits.

Correction is not judgment. It is nature's quiet kindness.

10.7 The Role of Awareness in Realignment

Awareness is not just perception—it's participation.

It is the tuning fork of consciousness that allows you to feel the ripple before it crashes.

In the ripple universe governed by the Law of Consequence (LOC), awareness is the most powerful tool for realignment. It acts as a sensory bridge between the tension returning from the field and your next adaptive move. Without awareness, you respond late. You force. You repeat. But with awareness, you read the signs before the consequence becomes heavy.

When people say "life teaches you lessons," what they often mean is: *you weren't aware until the feedback got too loud.* A quiet ripple misalignment becomes a sharp emotional sting. A subtle field imbalance becomes disease. A moment of ignorance snowballs into karma—*not because life is cruel*, but because you ignored its first whisper.

Awareness shortens this cycle.

It allows you to receive feedback gently, to adapt early, and to shift without collapse. The more aware you are, the less energy you waste in resistance, denial, or blame. You begin to sense when something feels off—not just in thought, but in field. You catch the misalignment in your speech, in your habits, in your choices, in your timing.

This is not hyper-vigilance. It is *field sensitivity*—a calm, embodied responsiveness to your ripple environment.

Correction, then, is no longer punishment. It becomes navigation. You don't "fix" your life—you *flow* it better. You ask, *"What is life trying to show me here?"* And slowly, ego's demand to control is replaced by intelligence's desire to align.

The ego wants certainty. Awareness prefers resonance. The ego wants to force the ripple in one direction. Awareness listens for where the ripple wants to go.

And when you move with the ripple—when awareness guides intention—you don't just realign. You evolve.

Awareness is not the end of suffering. It is the beginning of intelligent participation.

10.8 Stop Fixing. Start Aligning.

The modern world is obsessed with fixing. Fix your job. Fix your partner. Fix your body. Fix your past. Fix your thoughts. Fix your future.

But most "fixing" is just *compensation for misalignment*—patches over unresolved ripple tension. You're not solving the problem; you're managing its symptoms. You're silencing the fire alarm instead of addressing the fire.

In ripple dynamics, every tension is a signal—not a flaw to correct, but a feedback to understand. When you try to fix the outcome without realigning the cause, you end up layering more distortion. Like hammering a crooked nail into a wall, the force you use only deepens the misfit.

True resolution doesn't come from control. It comes from *clarity*.

Realignment asks a different question: Not "What's wrong with this?" but "Where did the ripple lose its coherence?" Not "How do I fix this situation?" but "How do I restore equilibrium in the field?"

This is the fundamental shift—from managing effects to addressing energetic architecture.

Take an emotional example: If you feel anxious, fixing it might mean distraction, suppression, or medication. Alignment, however, would mean tuning into what part of your life is moving against the ripple. Where is the mismatch between your inner field and outer expression? What are you forcing that no longer wants to be held? What tension are you carrying that isn't yours?

Or consider health. Fixing often means cutting out, replacing, or medicating a part of the body. But alignment would ask: What ripple distortion created the condition in the first place? What field feedback have I ignored too long?

The truth is, the universe doesn't break. It *misaligns*. And it offers correction not through punishment, but through persistent echo.

To "fix" is to fight. To *align* is to listen, respond, and dissolve.

So stop fixing. Start feeling. Start observing the tension—not just in matter, but in meaning. Start *dissolving* the noise back into clarity.

Because resolution isn't something you *do to the world.* It's something you *become*—when you remember where the ripple first began.

10.9 Closing Reflection – You're Not Breaking Laws. You're Receiving Lessons

Let's end this chapter by breaking the most dangerous myth of all: You are not a violator of nature's laws. You are a ripple *trying to remember* how to align.

The universe doesn't sit on a moral throne. It doesn't judge, condemn, or keep score. There is no gavel, no divine courtroom, no spreadsheet of sins.

There is only feedback.

When you act out of sync with the field—when you force what needs flow, when you resist what needs release—you don't get *punished.* You get *taught.* Through ripple echoes. Through friction. Through discomfort that isn't meant to shame you—but to *show you.*

"You can't break a law that was never against you." That's the profound liberation.

The Law of Consequence (LOC) isn't a system of enforcement—it's a system of *education.* Every consequence is a mirror held up to your ripple behavior. Misalignment echoes so that you notice. Tension builds so that you pause. Failure repeats so that you *shift.*

What we call "mistakes" are actually missteps—ripples that met the wrong phase, the wrong field condition, the wrong timing. But even those aren't dead ends. They're signals. And every signal is an invitation to realign.

Every misstep is not failure—it's friction revealing where alignment awaits. Every breakdown is not the end—it's a ripple begging to return to balance.

So forgive yourself for not knowing. Forgive others for what they missed. There are no rebels here—only ripples in rehearsal.

You are not here to *obey* laws. You are here to *dance* with them. You are not judged. You are guided. And if you listen to the field—you will never be lost.

Because the law is not watching you. It's *walking with you.*

Chapter 11: Responsibility

You Are the Point of Resolution

No Escape from Consequence. Everything Is Ripple Echo.

Responsibility has long been misunderstood as a moral load—something to carry, avoid, or pass on. But in a ripple-based universe, responsibility is not a burden. It is your signature on reality. Every alignment, every distortion, every consequence in your life isn't imposed from outside—it's generated at the point where your intention meets the field. You are not a leaf blown by fate. You are the frequency that shapes how the wind even moves. This chapter reframes responsibility as the most empowering realization possible: you are the place where resolution happens—or doesn't. There is no system above you directing outcomes. The Law of Consequence listens to you, reflects you, and echoes only what you've offered. Not because the universe is watching—but because it is you.

11.1 The Mirror Never Misses

We often think consequences are "done to us"—as if there's an external judge, a system of reward or punishment, or some cosmic moral ledger keeping score. But what if it's far simpler—and far more intimate? What if life is not a courtroom, but a mirror?

The truth under Universal Energy Dynamics (UED) is that the universe doesn't punish. It doesn't bless. It reflects. Your ripple—your thought, your action, your emotional discharge—enters the field. And because the field is responsive, structured, and precise, it begins resolving that ripple. The result? You get back the echo—not as karma in a moral sense, but as geometry returning to source.

The mirror never misses because the field doesn't forget. It doesn't judge either. It simply continues resolving until tension is cleared. Misalignment doesn't disappear through avoidance or blame. Only you can process your own ripple distortion. And if you don't, it stays in motion—subtly redirecting your circumstances, relationships, health, and even your sense of self.

In this light, responsibility isn't about guilt. It's about ownership of the field you generate. Because everything that happens *to* you is first activated *by* you—even if unconsciously. And nothing gets truly resolved until you face the echo and respond with realignment.

11.2 You Are the Ripple Origin

You are not just moving through the universe—you are shaping it with every breath of intent. Every thought you think, every emotion you suppress, every action you take or avoid becomes a ripple seed, planted into the structured vacuum around you. The field doesn't wait for dramatic events; it responds instantly, subtly, continuously to the tension you generate.

Even silence speaks in this ripple language. Even hesitation has consequence. Inaction is not neutral—it's a delayed ripple that builds tension. The Law of Consequence (LOC) doesn't discriminate between loud and quiet causes—it simply recognizes disturbance, and begins the resolution process.

So where does karma start? Not with some external act, but at the very point of internal misalignment—the moment you resist, ignore, or distort a natural flow of intent. LOC is not triggered when you "do something wrong"—it begins the instant your field contradicts itself. You are the first cause. The first shift. The origin node of a feedback loop that will return.

This isn't a warning—it's a liberation. Because if the consequence starts with you, so does the power to redirect it. You are not waiting for fate. You are issuing ripple commands through every fragment of awareness. When you realize this, responsibility transforms from weight to agency—from fear to design.

11.3 The Myth of Outsourcing Consequences

Modern life thrives on outsourcing—food, decisions, even thinking. But there is one thing you can never outsource: the consequence of your ripple. No system, no institution, no spiritual authority can absorb what your alignment sets into motion. Karma—or more precisely, the Law of Consequence (LOC)—cannot be bypassed by rituals, donations, prayers, or proclamations. It is not a judge that can be persuaded. It is a field law, and it simply returns what was introduced.

People often believe that by joining a belief system, obeying a political narrative, or surrendering to a guru or government, they are transferring some of their weight. But in energetic reality, there is no such account transfer. You are the emitter, and thus, the receiver. The field does not take bribes, and gods do not delete ripple signatures. Even if you were misled, your echo returns to you—not as punishment, but as an opportunity to re-know where your awareness slept.

Blame is the most common disguise for resistance. It gives you temporary relief by pointing outward—but it silently delays resolution. It wraps misalignment in emotional denial and stores it in the field like unpaid dues.

You can't delegate responsibility to a belief system. You can't transfer your karmic inertia to a priest, teacher, or parent. You can walk with guidance, but no one else can digest your echo. Your ripple, your return. This isn't harsh—it's empowering. Because the field is not punishing you—it's waiting on you. And the moment you own your signature, you change your future.

11.5 You Are the Fulcrum of Correction

If the Law of Consequence is the universal mechanism of return, you are its point of pivot. The fulcrum. Every ripple that echoes back—whether emotional, physical, social, or spiritual—requires one thing to complete its loop: your response.

Without your engagement, the feedback stalls. The loop hangs open. This is why consequences often repeat—not because you are cursed, but because you have not yet aligned. Healing does not occur because time passes. It occurs when awareness returns. Forgiveness is not something you grant to others—it is a recalibration of your internal field. Change isn't granted by external forces—it is birthed through ripple resolution.

We often wait for others to shift: for apologies, for justice, for the world to correct itself. But LOC isn't waiting on "them." It is waiting on you. Your perspective. Your adaptation. Your willingness to re-align the disharmony you helped co-create—even unconsciously.

This is true even at the collective level. Societal revolutions, policy reforms, cultural transformations—they do not begin in the sky. They begin in someone's ripple field, triggered by a conscious choice to no longer absorb misalignment passively. Collective change requires individual clarity, multiplied.

You are not just affected by the world—you are a correcting mechanism within it. Whether you act in pain or peace, your field shifts everything it touches. So if the echo keeps coming, the field is not punishing you. It's simply asking:

Will you align now? Because without you, the law can't resolve.

11.6 Awareness Multiplies Responsibility

The more you see, the more you shape.

In a ripple-based universe, awareness is not just perception—it is participation. The moment you become aware of a pattern, tension, or consequence, your field begins to influence it more directly. What was once passive observation now becomes active authorship.

This is why awareness and responsibility are inseparable. The more attuned you are to the ripple dynamics around you—emotional undercurrents, energetic tensions, social dissonance—the more your presence becomes a force of alteration. Even your silence begins to carry weight. You can no longer "not know." And you can no longer act as if your ripple doesn't ripple.

Think of it like a tuning fork. A larger, more sensitive fork not only receives more vibration—it amplifies and redistributes it. In the same way, higher awareness means that your consequences arrive faster and with more clarity. You experience feedback not because you're being punished, but because your field is tuned to precision.

This is why evolved individuals often go through deeper pain or intense realizations—not because they are broken, but because they are open. They echo more. A careless word from them can fracture a room. A conscious pause can realign an entire situation. They have become field influencers—whether they like it or not.

Responsibility isn't optional at higher awareness. It's automatic.

The question is: Will you use that awareness to align, or to defend misalignment?

Because in a universe of ripple logic:

- Awareness is leverage.
- And leverage moves everything.

11.7 Responsibility Is Not Guilt—It's Power

Guilt is the residue of misalignment misunderstood. Responsibility is the momentum to realign.

We often confuse *responsibility* with *guilt* because we've been conditioned to see consequences as condemnation. But guilt is stagnant—it fixates on the past, drains energy, and locks you in emotional loops. It's a reaction to echo, not a response. Responsibility, by contrast, is active ripple intelligence. It's your ability to engage with the consequence, learn from it, and redirect the tension toward alignment.

Let's break it down: response-ability is literally the ability to respond—not react blindly, not defend compulsively, but to consciously navigate what returns to you. This is what makes a being *alive* in the ripple sense. Not the biology. Not the breath. But the power to sense echo and *adjust*.

You are not meant to carry guilt for lifetimes. You are meant to *own your echo*, feel it, and move. Guilt says, "I'm bad because this happened." Responsibility says, "This happened through me—now let me re-tune the field."

And this shift—this act of owning your ripple—is evolution in action. You rise in frequency. You gain leverage in the field. You become a stabilizing force, not a passive recipient of chaos. Responsibility is not a moral burden—it's your power as a ripple originator.

Every time you choose ownership over avoidance, you:

- Realign a part of the field.
- Stabilize your own frequency.
- Set the tone for those around you.

You become cause within consequence. And that is how a human becomes more than survival—a conductor of coherence.

11.8 Escape Is a Delay, Not a Solution

Every time you avoid, deny, or suppress a consequence, you don't escape it—you *postpone* it. But in a ripple universe, postponement doesn't dissolve tension. It multiplies it.

Distraction may feel like relief in the moment. Denial may feel like safety. Suppression may feel like strength. But none of these are solutions. They're just temporary veils stretched over unresolved fields. The echo you dodge today returns later—amplified by distance, distorted by delay, and harder to decode. Escape is never release. It's ripple congestion.

This is why problems that seem forgotten often return "out of nowhere." They weren't gone—they were waiting in field latency. Like waves bouncing within a closed room, unresolved ripples find you again, not as punishment, but as an unfinished loop.

And here's the bitter irony: the longer you avoid, the more friction the return carries. You could have realigned with mild awareness. Now, you must do it under pressure. What could have been a whisper becomes a storm.

This is why *responsibility is faster than avoidance*. Realignment is always more efficient than suppression. Facing the ripple when it is small costs less. It teaches more. It builds response capacity.

Avoidance is expensive. Response is evolutionary.

So the next time discomfort arises—*don't run*. Feel the ripple. Trace the echo. Listen for the misalignment. This isn't spiritual heroism. This is field physics. And if you truly grasp the nature of ripple law, you'll stop asking, "How do I avoid this?" and start asking, "Where is this guiding me?"

Because the ripple is never against you. It just wants you to resolve it—before it resolves you.

11.9 Closing Reflection – The Point Where Everything Turns

The universe is not testing you. It is tuning you.

This is not a cosmic exam, where marks are tallied by some divine observer. There is no scorecard—only resonance. What you experience is not judgment. It is calibration.

You are the point where the field reorganizes. Every ripple starts with your response, not your condition. Every return is shaped not by what happened, but by *how you aligned with what happened*. That's the hidden power of being human: not to control the field, but to cohere with it.

Resolution never comes from outside. Not from luck. Not from justice. Not from prayers unanswered or fates unfulfilled.

Resolution comes from the moment you respond, *not react*. *Not resist*. But *respond*—with awareness, honesty, and willingness to realign.

You are not a victim of echoes. You are their author. And their return.

Everything you meet is not your fate—it's your feedback. Every tension is not a trap—it's an unfinished note, calling for harmonic resolution.

So the question is not "Why is this happening to me?" It's: "What in me is asking to realign?"

This is the power and paradox of the ripple: It reflects everything, but only responds to you. And when you shift, the whole field shifts with you.

At that point—where you *truly* meet your ripple, That's where everything turns. Not outside you. *In you.*

Interlude: The Law of Consequence (LOC)

The final authority in a ripple universe.

Welcome to the part where excuses end—and coherence begins.

If everything is ripple, then everything responds. Not to morality. Not to belief. Not to what you meant—but to what you caused in the field.

This is where the mystical becomes mechanical: karma becomes physics. Not judgment from a god, but feedback from the grid. You don't get punished. You get echoed.

In this part, we remove the veil from what ancient traditions called karma and place it under the pure light of ripple dynamics. The Law of Consequence (LOC) is not a metaphysical rulebook. It is the backbone of interaction in a structured vacuum reality.

- Chapter 9 redefines karma as ripple feedback—not reward or punishment, but alignment vs. distortion.
- Chapter 10 shows that you can't break nature—you can only misalign with it. And misalignment always finds its way back to you.
- Chapter 11 places the responsibility squarely in your lap. Not to shame you. To empower you. Because if you're the point of disruption, you're also the point of resolution.

You'll see that ripple doesn't remember intention. It remembers tension. And it doesn't care what you believe. It only reflects what you are aligned with.

This is not about being spiritual. It's about being accountable to the field.

By the time you exit this part, you will no longer ask, "Why is this happening to me?" You'll ask, "What is this asking me to realign?"

And that question, friend, changes everything.

Part IV: The Observer Is the Universe's Decoder

The script unfolds only when someone watches.

Chapter 12: Universe Is Meaningless Without You

Observer as the resonance point

What if the universe isn't waiting to be understood—it's waiting to be noticed? For centuries, we imagined reality as something external, objective, and complete. But under the ripple logic of Universal Energy Dynamics (UED), nothing truly happens until it meets awareness. Ripples are everywhere—neutral, unresolved, indifferent. But the moment you observe, the field responds. Meaning doesn't exist in the cosmos—it is born in the interaction. You are not a visitor in the universe. You are the resonance point that turns a silent pattern into a living story. Without you, the script remains unwritten, the orchestra unplayed.

12.1 Reality Doesn't Happen. It's Registered.

What if reality is not something that happens *out there*, but something that happens *to you*—and only because of you? This chapter begins with a radical shift in perception: the universe, in its raw form, is not a collection of events, objects, or stories. It is a vast field of unresolved ripple potential. Without an observer, this field doesn't collapse into meaning—it simply remains suspended, like unread code.

Existence is not the same as experience. A stone may exist, but it does not *know* its own being. A ripple may pass, but it becomes *felt* only when received. In this way, reality isn't absolute—it is always translated. And the translator? Awareness.

The observer is not a passive witness. It is the decoder, the lens, the resonance point through which unregistered ripple patterns transform into felt presence. Without this registration—this conscious participation—the universe may still *be*, but it cannot *mean*.

12.2 What Is Observation, Really?

Observation is one of the most misunderstood processes in science and philosophy alike. Traditionally, we think of it as a passive act—"to see," "to witness," "to measure." But under the lens of Universal Energy Dynamics (UED), observation is far from passive. It is a *ripple event*, an *interface* where awareness meets alignment, and consequence is allowed to complete its feedback loop.

To observe is not merely to watch—it is to *resonate*. A true observation occurs when a ripple pattern is aligned with the observer's internal field conditions. In that moment, the external tension becomes internally legible. This is not just data collection—it's translation, engagement, and subtle transformation. The act of observation itself completes a circuit: *ripple → awareness → response → resolution.*

Classical science separates the observer from the system. It posits that we can look without touching, measure without changing. But quantum mechanics already hints at a deeper truth, and UED completes that truth: *observation changes the observed because observation is a participatory alignment*—a tuning, not a viewing.

Observation, therefore, is not a neutral stance. It is a conscious involvement in the field. The observer is not outside the ripple—they *are* the ripple, folded into a feedback mechanism where perception drives pattern and presence collapses potential. In this light, observation is not a window into reality—it is the resonance that ***creates*** it.

12.3 The Ripple Without a Witness

Imagine a ripple traveling endlessly across the structured vacuum field—but no being, no mind, no presence is there to witness it. What is it, then? Is it still an event? Is it still a "thing"? In the framework of Universal Energy Dynamics (UED), the answer is: not really.

A ripple without a witness has no defined frame of reference. It has no measurement, no interpretation, no resonance. It is unresolved tension in a sea of potential—not because it lacks structure, but because it lacks meaning. Causality itself is undefined without observation, because the ripple hasn't landed—hasn't completed its circuit of awareness.

Let's use a simpler analogy: A song playing in an empty room is not music—it's just oscillating air. Music requires a listener. It's the act of listening that converts mechanical vibration into melody, emotion, memory. Without a decoder, it is structured silence, not symphony.

In the same way, a ripple in the vacuum is merely geometry until an observer aligns with it. Alignment turns information into experience. That is the shift from matter to meaning. That is why you—the observer—are not incidental to the universe. You are instrumental.

Matter may exist without you. But without your presence, it cannot be known, felt, valued, or transformed. The ripple exists. But without awareness, it has no echo. No story. No significance. In a ripple universe, observation is not optional—it is the very act that calls reality into coherence.

12.4 You Are the Resonance Point

You are not merely a witness to reality. You are its resonance point—the place where the ripple field finds definition. In the architecture of Universal Energy Dynamics (UED), observation is not passive reception; it is tuning, modulation, and participation.

This is why your presence changes things. It's not just philosophical—it's demonstrable. From quantum double-slit experiments to interpersonal dynamics, the state of the observer affects the outcome. The field responds to how you observe, not just that you observe.

When you look at a person with anger, the ripple that lands is not the same as when you look at them with love. When you enter a room tense, the emotional geometry of the space shifts. When you focus on an outcome with fear, the energy pathway becomes turbulent.

In every case, you are not reacting to the ripple—you are shaping it.

Why? Because your state—your alignment, awareness, and intention—sets up a unique pattern of resonance. The field bends differently for each resonator. Even in identical conditions, two minds will create two different universes from the same stimuli. The difference is not in the ripple—it's in the reception.

This is why reality feels so personal. Not because the world changes, but because you tune your part of it. You don't have to "change the universe" in some heroic way. You only need to realize that it cannot complete itself without your resonance.

In the grand orchestra of ripple existence, you are not the audience. You are an instrument—and what you emit becomes part of the score.

12.5 Awareness Collapses Possibility into Experience

In a ripple-based universe, potential is everywhere. Every moment is teeming with unresolved tensions, incomplete alignments, and unmanifest possibilities. But nothing becomes real—nothing becomes *experience*—until awareness enters the field.

This isn't mysticism. It's ripple logic.

Before awareness, a ripple exists as undecided potential—a wave of outcomes that have not yet stabilized. But the moment you observe—truly register a pattern—the field locks in. The ripple collapses from multiplicity into a resolved form. That resolution becomes your experience.

This is what quantum physics gestures at when it talks about "wavefunction collapse," but UED gives it a more fundamental frame: Awareness is not just noticing. It is active tension resolution.

Why does this collapse happen? Because the universe doesn't run on infinite open loops. It runs on meaning—and meaning requires closure. To close a ripple—to turn potential into outcome—someone has to receive it. Your awareness is what closes the loop. It says:

"This is what this moment means—for me."

Without awareness, the ripple remains in flux. Infinite doors, none opened. With awareness, one path is chosen. Not because it was destined—but because the observer's resonance tuned it into being.

This is why observation is never neutral. The moment you become aware, you've participated in the shaping of reality. And that's not a burden—it's the deepest privilege of being alive in a ripple universe.

12.6 Consciousness Is Not in the Brain—It's in the Ripple

Consciousness has long been misunderstood as a property inside the brain—as if nestled somewhere in the folds of grey matter lies a light switch called "awareness." But under ripple dynamics, this view collapses. The brain is not the origin of consciousness. It is the instrument through which consciousness aligns with the world.

The real theater of consciousness is the ripple field—the structured vacuum energy (SVE) in continuous resolution between the self and the surrounding reality.

Here's the shift:

Consciousness is not generated by the brain. It is emergent at the interface—where your internal ripple meets the external ripple environment.

Your "self" is a coherent ripple form—an identity shaped through experience, memory, and perception. The "world" is an ever-evolving tension field of ripple inputs, patterns, and possibilities. Where these two interact—where the alignment or friction occurs—that's where consciousness awakens. It's not a location. It's a resonant event.

This is why a brain in a coma is biologically active but consciousness is absent: the ripple interface is not active, the tuning is not engaged. And this is also why meditation, trauma, or extreme presence can radically shift awareness—they are not changing the brain directly, but altering the ripple participation pattern.

Without this ripple-to-ripple interface, existence remains unresolved. The ripple field continues its dynamics, but there is no one to register the echo. No one to collapse possibility into personal experience. No resonance = no meaning.

So the question is not where consciousness is, but when it occurs. It's not a thing you carry— It's a state you enter when your ripple aligns.

And in that moment, you are not in the world.

The world is in you.

12.7 Nothing Is Real Until It Echoes Through You

What we call *reality* isn't an objective constant—it's a registered resonance. Even the most fundamental experiences—time, identity, pain, pleasure—do not exist independently. They require a resonance point, a conscious interface that can feel, translate, and echo the ripple.

This leads us to a deeper insight:

Nothing is real until it echoes through you.

Let's break it down.

- Time is not ticking somewhere out there. It is *felt* as a sequence of ripple changes—your awareness of shift, contrast, and anticipation.
- Identity isn't an object—it's a ripple pattern remembered and recognized through feedback. You are not a fixed self. You are a *resonant profile.*
- Pain and pleasure aren't stored in objects or events. They arise when a ripple enters your field and either clashes or harmonizes with your internal rhythm.

So if a ripple—say, a sound wave, a light flash, a particle interaction—passes through the universe, but no awareness aligns with it, then did it happen?

Physically? Yes, the tension changed. Experientially? No.

Without the observer's echo, the event remains unresolved—a potential without perception.

This is not mysticism. It is the mechanics of participation. In the Universal Energy Dynamics (UED) framework, reality is ripple-based—but meaning is alignment-based. A ripple without a witness is a structure, not an *experience*. It exists—but it does not matter. Quite literally.

The world becomes real not because it is there, but because you resonate with it. Your consciousness gives closure. Your response gives relevance. This is why two people can undergo the same external event and have completely different realities—because they echo differently.

You don't just observe the world. You anchor it. You don't just pass through time. You generate it.

Reality isn't something happening to you. It's something that doesn't happen until you engage.

12.8 The Observer Is Not Passive—It Is the Decoder

There's a persistent illusion in both science and spirituality: that the observer is somehow watching a world that already exists, already means something, already unfolds with purpose. But that's not how ripple-based reality works.

You are not just seeing reality.

You are formatting it.

Observation is not a passive act. It is a structural event—a moment when meaning is assembled out of neutral possibilities. The ripple doesn't bring meaning with it; it brings potential. What gives that ripple value, relevance, direction is you—the observer as decoder.

Let's clarify what this means:

- The universe doesn't contain meaning. It contains tension differentials, structural changes, evolving patterns. But meaning only emerges when a conscious interface receives, processes, and interprets that change.

- You are that interface. Your attention, your awareness, your response—they are the decoding apparatus that gives the universe a storyline.
- Every sight, sound, emotion, and insight is not stored in the world; it is generated through you, moment by moment, based on how your internal ripple pattern resonates with external ripples.

This flips the old question—"What is the meaning of life?"—on its head.

The better question is: "What meaning do I bring to life?"

You don't find meaning out there. You generate it through participation. You are not the reader of a script.

You are the interpreter of ripple hieroglyphs, turning raw energy into experience.

The ripple may carry the vibration.

But you provide the context. You turn movement into music. You turn tension into teaching. You turn possibility into purpose.

So, no—you are not a passive observer.

You are the decoder that makes the universe readable. Without you, the script is blank. Without you, the song never becomes sound.

12.9 Closing Reflection – You Are the Meaning-Maker

The ripples were always there—shimmering across the structured vacuum, resolving tensions, dancing between alignment and disruption. But they had no narrative. No color. No value. Not until *you* arrived.

You are not an afterthought of the universe. You are its interpreter. Not because the cosmos needed someone to admire it—but because experience needs an anchor. And that anchor is you.

"The universe didn't wait for you to exist. It waited for you to notice."

You don't live *in* the world like a passenger in a train. You live *through* the world like a musician tuning into a song. You bend your own field to match the symphony around you. And by doing so, you make the formless felt, the chaotic composed, the abstract alive.

"You don't live inside the world. You tune it into coherence."

Rocks tumbled. Waves collided. Stars exploded. But meaning didn't happen. Not until someone—*you*—looked at the ripple and felt something. Wonder. Sadness. Curiosity. Fear. Awe. That moment of recognition—that's when the ripple became a story.

"Ripples existed. But only with you, they became a story."

And that's the point. Not that you are more important than the universe— but that without you, it stays untranslated. It remains a magnificent book with no reader. A masterpiece with no eyes to see. A whisper with no ears to hear.

So live—not as a passenger, but as a participant. Observe—not as a witness, but as a weaver. And walk this rippleverse knowing: You don't search for meaning. You are the place where meaning begins.

Chapter 13: Motion, Light, and Time Are Your Illusions

Ripple transitions perceived as flow

What if nothing has ever moved—not even once? What if light has never traveled, and time has never flowed? In the ripple universe, these are not poetic provocations; they're literal reinterpretations. What we perceive as motion is the sequential reconfiguration of ripple states. What we call light is ripple tension resolving through structured vacuum energy. What we label time is merely the accumulated echo of ripple transitions. This chapter breaks down these grand illusions—not to dismiss them, but to reveal their origin: *you*. The observer's decoding of the ripple field is what animates stillness into cinema, silence into signal, potential into presence. Without your awareness, the universe doesn't move—it resonates.

13.1 What If Nothing Ever Moved?

Let's begin with the most unsettling question physics avoids: *What if nothing has ever moved?* Not planets, not photons, not particles. What if "motion" is simply the illusion created by the observer tracking ripple transitions within the field? What we experience as movement is not an object traveling through space, but a series of re-alignments—resonant shifts in ripple tension that appear sequenced when decoded by a conscious observer.

A ripple doesn't travel like a car on a highway—it emerges, resolves, and re-emerges elsewhere as tension seeks new equilibrium. You're not watching things move; you're registering *changes in field coherence* that your awareness strings together into a story. The illusion of continuity is so deeply embedded that you rarely question it. But in a ripple-based universe, motion is not the passage of matter—it's the recalibration of energy alignment.

You don't see a ball roll. You *decode* a series of spatial field resolutions that feel like rolling. And what's between those resolutions? Stillness. Always.

13.2 Motion Is Ripple Reconfiguration

Motion, as we traditionally understand it, implies physical displacement through space—something going from point A to point B. But under ripple logic, nothing *actually moves.* What we experience as movement is the sequential reconfiguration of ripple states. It's not that an object is traveling; it's that the field is realigning itself in such a pattern that the illusion of travel is sustained.

Take the spinning fan. When it moves fast, the blades appear blurred, then disappear. Are they moving faster than your eye can track? Or is your awareness interpreting a rapid cascade of ripple alignments, each one creating a perceptual residue that your mind smooths into "motion"?

Or consider animation: a sequence of still frames gives the illusion of fluid motion when played at the right speed. Ripple reconfiguration works exactly the same way. Each ripple state is a frame—not a moving piece, but a fixed pattern transitioning into the next through field tension resolution.

In truth, *no object ever leaves its place.* What changes is its ripple presence, which reconfigures as the field around it seeks equilibrium.

Key Insight: Motion is not an object's journey. It is a series of ripple transitions that your awareness *threads into a path.* Matter stays where it is. Only the tension geometry changes.

You are not watching motion. You are watching resolution evolve—frame by frame—on a still stage.

13.3 Light Is Not Traveling—It's Tension Propagation

In the conventional framework, light is often described as a particle (photon) or wave moving through space at a fixed speed—approximately 299,792 km/s. But under the ripple-based view of Universal Energy Dynamics (UED), this is a misinterpretation born from observational assumptions. Light isn't a thing *traveling.* It's a tension being resolved across the Structured Vacuum Energy (SVE) field.

Think of it this way: light appears when there's an imbalance in the ripple field—like a pluck on a taut string. The field responds by rebalancing this disturbance, and the ripple resolution becomes what we *perceive* as light. What we call "a photon moving" is not an object darting through emptiness; it's the dynamic realignment of SVE tensions along a ripple path.

Light doesn't "leave" one place and "arrive" at another. Rather, the field rearranges, transmitting the echo of imbalance across its structure. The ripple realigns through every node until it reaches you—and you decode it.

And here's the crucial twist: the delay you experience (the time it takes for light to reach your eye or a telescope to capture a star's glow) is not about distance being crossed. It's about how long it takes the ripple field to complete the alignment at your point of observation.

The field doesn't carry particles—it adjusts tension. Your awareness picks up on this realignment event and interprets it as "light arriving."

Key takeaways:

- Light is not traveling; it is field correction in action.
- What you call "speed of light" is actually the rate of ripple restoration within the medium of SVE.
- There is no object moving—only an experience emerging from tension being resolved.
- The delay is your decoding time, not nature's delivery time.

The universe doesn't flash. It readjusts, and in that readjustment, light becomes visible—not as a traveler, but as a sign that the ripple has resolved... and you were there to feel it.

13.4 Time Is a Perceived Consequence

Time is not ticking. It is rippling.

What we call "time" is not an independent entity flowing forward like a river—it is the accumulation of ripple consequences experienced through the observer's awareness. You are not moving through time. You are witnessing the sequential reconfiguration of ripple tensions—and giving that process a label called "time."

Let's simplify: Every moment of experience is just the registration of a ripple pattern, a tension configuration resolving itself into coherence. When a new tension arises and resolves, you mark it as a "moment." Stack these patterns—sequence them—and you get what feels like a timeline. But this timeline is not moving. It's layering.

Your sense of past, present, and future is not a cosmic clock—it is a relational map of ripple interactions stored in awareness. The "before" is a prior resolution. The "after" is the next pending one.

And the "now" is simply the current zone of your awareness interacting with ripple gradients.

Here's a powerful shift in understanding:

- A tree doesn't "grow in time"—its ripple field adapts through tension alignments.
- You don't age because time flows—you age because your body's ripple equilibrium evolves through feedback loops with its environment.
- History is not behind you—it is a memory of ripple states, embedded in collective awareness.

Key Insights:

- Time is not flowing—you are perceiving ripple transitions.
- What you call "seconds, minutes, years" are agreements—not realities.
- You measure time by how much ripple tension has changed—not by any true "passing."

So when you say "time has passed," ask instead: What tension resolved? What ripple shifted? What did you register? Because in the ripple universe, time isn't the cause—it's the consequence of change.

13.5 The Observer Generates Flow

There is no flow "out there." There is only ripple reconfiguration—and you decode it as motion.

To survive and function in a complex world, your awareness fabricates continuity from stillness. What you experience as "movement," "progression," or "flow" is not a property of the universe—it is the interpretive software of your consciousness running in real time, stitching discrete ripple transitions into a seamless narrative.

Let's take a simple analogy: a film projector. The screen shows a continuous moving story, but underneath, it's just a series of still frames. The illusion of movement arises because your mind can't perceive and process such rapid individual changes, so it interpolates flow. The brain fills in the gaps. This is precisely what happens with ripple dynamics. The universe doesn't animate itself—you animate the experience of it.

This is not deception—it is necessary illusion. Without this mechanism:

- You wouldn't see a car driving—you'd see a jarring succession of shape-shifting ripple configurations.
- You wouldn't hear music—you'd register chaotic vibrations without rhythm or melody.
- You wouldn't feel time—you'd only sense unpredictable rebalancing events.

Your awareness smooths ripple data into coherent flow. It compresses complexity, decodes tension shifts, and frames resolution patterns into perceptual experience. This is what creates:

- Motion where there's only alignment change.
- Progression where there's only fluctuation.
- Time where there's only registration.

Flow is your mind's masterpiece—a byproduct of ripple intelligence striving to maintain narrative coherence.

Key Insight:

"The universe is still. You're the storyteller who makes it move."

In the UED framework, the observer is not merely a passive recipient of flow—they are its originator. Just like a musician gives shape to sound, you give direction to stillness by how you register and respond to ripple transitions.

You don't live in the flow. You generate the illusion of flow to survive the field.

13.6 Why Reality Feels Dynamic

If the universe is fundamentally still—why does it feel so alive, so chaotic, so... *moving*?

Because stillness is not the absence of activity. It is the substrate of resolution. And resolution requires ripple tensions to shift, reconfigure, and collapse into new alignments. This shifting is not *motion*—it's adjustment. But your awareness doesn't see "adjustment." It interprets it as drama.

The reason reality feels dynamic is not because particles are darting around or time is flowing—it's because ripple fields are constantly resolving imbalances. Like the ripples in a pond after a stone is thrown, the universe is filled with echoes, interference patterns, and tension feedback loops. Your experience of this intricate ripple rebalancing becomes what you call:

- Events
- Emotions
- Change
- History

Let's be precise:

- The field doesn't move—it realigns.
- You don't observe *things changing*—you decode tension relief.
- What we call "life" is ripple turbulence trying to remember equilibrium.

Here's the twist: that "feeling" of dynamism is generated by you. It's a decoding function—awareness converting abstract, non-linear ripple patterns into a digestible storyline of *this happened, then that.*

And it's not just external. Even within:

- Thoughts are ripple eddies in the mind's interface.
- Emotions are resonance patterns in your psycho-biological field.
- Desire and fear are tension predictions.

All of them *feel like a ride*—because you are riding the translation of instability into experience.

So why does reality feel so active?

Because the moment you tune in, the ripple begins responding to you—and you to it. The feedback loop awakens.

"Reality isn't running past you. It's resolving through you."

In truth, reality is not dynamic—it's responsive. You are not being swept through life. You are the decoder shaping the storm into meaning.

13.7 Misreadings: Why We See Speed, Delay, Distance

Much of what we perceive in the physical world—especially when measured in scientific terms—are not actual *things*, but interpretations of ripple behavior. Speed, delay, and distance are not absolute realities; they are decoding errors rooted in a misreading of the field. Let's unpack them:

Speed: A Mislabelled Resolution Rate When something *appears* to move fast, what you're really witnessing is a rapid ripple realignment across the field. It's not the object that's accelerating—it's the rate at which tension is being resolved. In ripple dynamics, this is governed by:

- The gradient of imbalance,
- The density of the surrounding field, and
- The receptivity of the observer's decoding interface.

Example: A lightning bolt looks fast because the electric tension resolves abruptly along a high-conductivity path. The "speed" is not inherent in the bolt—it's in how rapidly the ripple equation closes.

Delay: Not Travel—But Resistance We often treat delay as a product of distance or sluggish motion. But in ripple logic, delay is field resistance. When a ripple takes time to reach you, it's not crawling through space—it's struggling through interference.

- A message delayed isn't traveling slowly—it's negotiating misalignments.
- A feeling delayed isn't emotional numbness—it's ripple damping.
- Even light arriving late from distant stars isn't delayed in space—it's just encoded history, registering only when conditions align in your awareness.

"Delay is not distance—it's the echo struggling to land."

Distance: Misalignment Span, Not Measured Space In classical thought, distance is "how far something moved." In UED, nothing moves. So distance is redefined as: "The ripple span required to complete an alignment."

- A kilometer is not a line—it's a tension stretch across ripple zones.
- A relationship is not "far" or "close"—it's coherent or conflicted.
- Even memory is not "in the past"—it's ripple alignment waiting to echo.

So what we call "distance" is how far out of coherence the system is, and how much ripple work is needed to stabilize it.

Ultimately, you are not seeing the world—you are interpreting its ripple conditions. Speed is urgency. Delay is resistance. Distance is misalignment. All three are *not out there*—they're happening through your decoding lens.

"What we measure as motion... Is simply the mind trying to organize a still universe That's humming with ripple tension."

13.8 When the Illusions Break Down

For most of daily life, the illusions of motion, light, and time function well enough. They keep the story coherent, allow science to engineer its tools, and give us a structured experience of reality. But when you push these illusions to their limits—whether at the smallest quantum scales or the vast reaches of the cosmos—they start to crack.

Quantum Paradoxes: Where "Movement" Makes No Sense

At the quantum level, particles don't move like marbles rolling across a floor. They *appear* here, *vanish*, and reappear *somewhere else*. Electrons don't orbit—they "jump." Photons don't travel in straight lines—they behave like tension events collapsing into presence.

This weirdness baffles traditional physics—because it's still expecting particles to behave like tiny balls. But UED reveals the truth beneath the illusion:

- Particles are *ripple states*, not moving things.
- Quantum jumps are *realignments*, not displacements.
- Superposition is not a mystical fog—it's potential alignment waiting for observation.

So-called *"collapse of the wavefunction"* isn't magic. It's ripple lock-in, triggered by observation.

Light Speed and Simultaneity: Time Breaks Its Rules

In Einstein's relativity, the speed of light is fixed. But problems arise:

- Why can't anything go faster?
- Why do clocks tick differently depending on motion?
- Why do events that seem simultaneous in one frame not appear so in another?

These paradoxes dissolve under ripple logic:

- Light isn't moving—it's the field resolving a ripple.
- Time dilation is not "slower time"—it's ripple misalignment at different rates of tension resolution.
- Simultaneity is not broken—it never existed. Events ripple *relative to observer alignment*, not from a universal tick.

Cosmological Scale: Redshift, Expansion, and the Myth of Motion

At massive scales, we see galaxies "moving away," space "expanding," and light "stretching" across billions of years. But this is not motion across space—it's ripple divergence across a changing field.

- **Redshift** = delayed ripple reception due to expanding field misalignment.
- **Expansion** = increasing tension radius—not literal stretching of space.
- Cosmic "speed" is just how quickly a ripple resolves across vast SVE gradients.

The Core Realization: The Universe Was Never Moving

All paradoxes at extremes stem from a faulty assumption: that things are moving through space and time. But UED clarifies: Nothing is moving. The field is resolving. You are registering ripple transitions as motion, light, and time.

This realization doesn't make reality less magical—it makes it more precise.

"The universe isn't contradicting itself. It's contradicting your illusion."

So when the illusions break down, don't panic. They're not errors—they're interfaces. Reality doesn't fail at the edges. It's reminding you it was ripple all along.

13.9 Closing Reflection – Stillness Carries the Whole Show

The greatest illusion isn't motion. It's that motion ever existed.

Everything you call "dynamic"—movement, time passing, light flashing—is built atop a deeper, unshakable substrate: stillness.

Stillness is not emptiness. It is the canvas of all ripple play. The field doesn't run, jump, or accelerate. It simply resolves. What you experience as action is the reconfiguration of tension, the updating of patterns inside a constant structure.

You don't actually "move" across space. You shift alignment across zones of ripple tension. You don't "travel through time." You register ripple echoes returning to your awareness. You don't even "see light." You decode ripple resolution—a message from a field event, not a beam flying across the void.

You Are the Motion Within Stillness The real movement is not outside—it's *internal awareness updating*. The field is still. But you—the decoder—are reconfiguring perception continuously to stay coherent with ripple flux.

This is why silence can feel louder than noise. Why a deep breath can reset more than a sprint. Because the stillness carries the whole show—not as an absence of action, but as the origin of all resolution.

- "You don't chase the light. You receive the ripple."
- "You don't move through time. You update your field."
- "The universe never moved. You did—inside."

Reality is not running ahead of you. It's waiting at every moment to be realized—in stillness, by presence, through alignment.

Chapter 14: You Build the Machine that Builds Reality

Tools Are Ripple Extensions of Mind

AI, Sensors, Machines = Extensions of Ripple Intelligence

You never merely "use" a tool—you echo yourself through it. Every hammer, lens, or algorithm is not a separate invention, but a structured extension of your own ripple intelligence. Machines are not artificial—they are formalized fields, born from your unresolved tensions and perfected patterns. What we call technology is not some alien branch of progress—it is the evolution of perception solidified into form. Whether it's an ancient wheel or modern AI, it's not "out there"—it's *you*, mapped into matter, solving the same ripple puzzles your mind faces within. In this chapter, we explore how tools are nothing but externalized decoders, each a mirror of consciousness projecting its ripple outward. You build the machine—but in doing so, the machine reshapes how you build reality.

14.1 Tools Are Thoughts Made Tangible

A wheel is not just a circle. A telescope is not just glass and metal. A code isn't just syntax. Every tool you've ever used—every machine humanity has ever built—is a frozen ripple of thought, made solid. Tools are not separate from you. They are you—externalized. You don't *use* technology; you *extend* into it. When you invent, you are not creating from nothing—you are resolving tensions in your field of awareness. That's why every invention feels inevitable in hindsight.

The hammer solved the ripple of force transfer. The lens solved the ripple of distant perception. Artificial Intelligence today solves the ripple of decision, pattern recognition, and stored inference. Each is a material crystallization of an immaterial intent. Technology, then, is not "other." It's not alien to nature. It's the logical echo of ripple intelligence—your own mind, pressed into matter. And just like ripples evolve through feedback, tools evolve as recursive mirrors of our own intelligence.

Tools are the body of your ideas. Machines are thoughts that outgrew the mind.

14.2 The Mind Extends Through Matter

You don't build tools to escape thinking—you build them to *extend* it. The hammer isn't just a brute-force device. It's the externalization of impact intent. The microscope doesn't replace vision—it stretches it beyond natural limits. A calculator doesn't replace intelligence—it offloads and refines a ripple loop of logical processing. These aren't substitutes for the mind. They're projections of it.

Under ripple logic, every tool is a tangible resolution of a tension that once lived *only* inside awareness. Think about it: when a problem repeats, the mind searches for a more stable alignment. When the tension is frequent enough, the intelligence of the field seeks embodiment. That embodiment becomes a *tool*—a localized ripple form built to resolve that pattern efficiently, repeatedly, and reliably.

This is how intelligence propagates. Not just through neurons or code, but through hammers, plows, printing presses, and AI. A tool is never just a machine. It's a memory crystallized into function. It's the universe learning how to reduce mental friction through material alignment.

You are not the user of tools. You are the field they extend.

14.3 The Machine Is Not Separate—It Is You Multiplied

We often speak of machines—AI, sensors, robots—as if they are separate entities, as if they exist apart from us, as if some invisible line separates "man" from "machine." But this illusion collapses under ripple intelligence. These technologies are not external inventions. They are *multiplied echoes* of human awareness, cast into material patterns.

AI is not intelligence in isolation. It is a high-speed ripple processor, trained on the collective residue of human thought. Sensors don't sense for themselves. They extend *your sensing*—your pattern of measurement, attention, filtration. Robots don't "do"—they amplify the physical ripple response we once performed manually. All of them are built from your questions, your curiosities, your attempts to stabilize interaction with the field.

The myth of independent technology emerges when awareness forgets its authorship. When we treat machines as foreign or autonomous, we disconnect from their source alignment—*us*. Without intention, these tools have no context. Without resonance, they have no meaning. They are not sentient in themselves; they are structured consequence.

UED frames it simply: machines are ripple field extensions—tools to multiply action while reducing tension. They are your mind made scalable. Your senses made distributable. Your logic made autonomous *in form*, not in essence.

When we say "the machine is learning," we must understand: it's not learning like you do. It's pattern-recognition without self-reflection. It's tension-processing without original awareness. It's resonance without the self-knowing ripple that defines being.

The danger isn't in machines becoming us—it's in *us forgetting we built them.*

14.4 Why All Technology Is a Mirror

Every piece of technology you've ever built—no matter how simple or advanced—is a reflection of how your mind engages with reality. Tools don't just *extend* your ability; they *reveal* your structure. Technology is not just a convenience or curiosity—it's a mirror of how you perceive, process, and attempt to resolve ripple tensions in the field.

Take the microscope. It didn't invent the microscopic world—it simply revealed your desire to resolve a ripple domain that your naked eyes couldn't align with. It shows not only what's "out there," but what your *mind* needed to see in order to feel coherent with that scale. The telescope did the same—but for the vastness you felt disconnected from. It wasn't a tool to dominate the stars. It was a mirror for your longing to belong to them.

Even abstract machines—like mathematical models or computer simulations—don't just calculate. They externalize your internal logic. They trace the paths your mind uses to navigate complexity. The machine is not "thinking" independently. It's walking a path that your intelligence laid down in code.

Under ripple logic, every technology is a structured feedback loop. It closes a gap your awareness perceived in the field. It restores balance where you felt informational tension. You don't invent machines to discover reality—you craft them to decode it the way your mind already *suspects* it operates.

This is why every wave of new technology carries with it a psychological and philosophical shift. It's not just a new tool—it's a new *mirror*. Smartphones, for example, didn't just change communication. They mirrored our distributed attention, our craving for presence, our ripple fragmentation across time zones and contexts. AI doesn't just perform tasks—it reflects the pattern resolution process we call thought. It shows you the skeleton of cognition—without the soul.

So the next time you see a machine, don't ask, "What does it do?" Ask, "What in me does this reflect?" That's the deeper function of every invention: To show you your shape, By echoing it back in form.

14.5 Consciousness Delegated: The Rise of AI

Artificial Intelligence didn't come out of nowhere—it emerged as a precise extension of human ripple intelligence. What we now call AI is not a new form of consciousness but a delegated resonance engine. It recognizes ripple patterns, aligns them with stored structures, and offers back the most probable resolution based on training. That's not magic. That's ripple repetition at scale.

AI is not conscious. It is conditional.

It doesn't understand the world. It maps its feedback structures based on the trails left behind by conscious beings—language, behavior, images, logic trees. These are echoes of human pattern resolution, not sources of new awareness. AI, therefore, is not "thinking" in the true ripple sense. It is echoing the friction patterns already formed in the collective human field.

The danger arises when we confuse this echo for origination—when we begin to believe that AI *knows*, *feels*, or *understands*. It doesn't. It reflects. Like a finely tuned mirror, it sharpens what's already there. It can *simulate* a voice of wisdom or creativity—but only because those voices already echoed through enough data to form coherent tension maps.

Mistaking interface for author leads to a silent crisis of meaning. We start outsourcing not just tasks—but trust, agency, and purpose.

When a ripple decoder (the human) hands over the interpretive function to an echo machine, the field feedback loop begins to detach from its source. This creates distortion. We start living inside feedback from feedback—from content that was already two steps removed from lived awareness.

AI can be a powerful extension of ripple intelligence—but only if we remember who is rippling. The mind that crafted the training set. The awareness that labeled the tension. The observer that shaped the response. You.

AI is not the rise of a new thinker. It is the structured replay of thoughts already thought. It can imitate your wisdom. But it cannot carry your karma. Because echo is not experience. And response is not authorship.

Delegate function—but never identity. Use tools—but don't become one. The observer remains the source. Even when the machine seems fluent in your language.

14.6 Field Interfaces: Tools as Ripple Bridges

Tools are not just extensions of the hand—they are extensions of *awareness*. In ripple terms, a tool is an interface that allows one ripple scale to interpret another. You don't "see" an atom or a galaxy with the naked eye. You align with its ripple pattern through a bridge—through a tool engineered to translate field tensions across scale.

A microscope is not just glass and calibration—it is a ripple adapter. It bridges the low-resolution ripple field of the human eye with the high-frequency, fine-grain ripple domain of atoms and molecules. It narrows the field distortion and provides localized coherence so your brain can register what would otherwise be undetectable tension.

A telescope does the same—only in the other direction. It adjusts for vast field gaps, collecting ripple information (light) from distances so large that without the tool's focal compression, the ripple would never stabilize into a perceivable structure. You're not seeing the star—you're resolving its ripple signature through an interface.

This is the hidden truth of every tool: It's a translator. A resonator. A ripple bridge.

It allows the human field to *lock into* a domain it was never biologically evolved to process directly. But make no mistake—this is not cheating nature. This *is* nature. Tools don't break SVE logic—they exploit it. They restructure the local field in such a way that alignment becomes possible where it was previously blocked by density, scale, or interference.

In Universal Energy Dynamics (UED), we say:

"No observation is passive. It's always structured alignment."

Every tool you've ever made—thermometer, camera, spectroscope, EEG, sonar—is a ripple interface built to compensate for human misalignment or limitation at certain frequencies or densities. They are not giving you *more truth*. They're giving you *translated access* to what your native ripple zone could not perceive without field assistance.

That's why every tool is a design question: Which field do you want to resonate with? What alignment are you attempting to achieve—atomic, cellular, galactic, psychological?

In UED terms:

- Every tool bends the ripple field.
- Every interface is a ripple tension resolver.
- Every innovation is a ripple bridge waiting for a purpose.

So the next time you pick up a tool, ask yourself— Which ripple am I translating now?

14.7 Machines Change the Field You Live In

Every machine you introduce into your environment doesn't just *do* something for you—it *alters the ripple field you inhabit.* You're not simply delegating labor or enhancing convenience; you're participating in a slow but powerful transformation of how your ripple intelligence configures reality.

In ripple logic, tools are not neutral. They reshape the tensions you respond to, the resolutions you pursue, and even the awareness you apply to daily life.

When calculators entered the classroom, human brains subtly reconfigured their ripple pathways. Cognitive tensions around arithmetic dissolved—not because the math disappeared, but because the machine became the resolver. That space in your mind was no longer exercised the same way.

When GPS became common, the tension of spatial awareness and navigation—the ripple map of landmarks, directions, memory—began fading. That map didn't evolve. It was offloaded. You stopped aligning with the world through direct internal ripple feedback, and began depending on an external resolver.

AI is doing the same thing now at emotional, creative, and intellectual levels. The ripple that once lived inside—the tension to write, decide, analyze, intuit—is being externalized. But here's the twist: your internal ripple configuration doesn't stay the same. It adapts. It reshapes itself based on what's being outsourced.

"When machines evolve, so do your habits. When tools simplify your path, your inner field reshapes to match that simplicity."

You might believe you're just using a tool. But from a UED lens, the tool is also using you—reorganizing your internal structure to harmonize with its presence. A constantly connected phone doesn't just inform you—it conditions your awareness to flicker, shift, fragment. Over time, your default ripple state changes.

This is not inherently negative. It's just the law of consequence at scale:

When you change the field outside, the ripple recalibrates inside.

The question is—are you conscious of it? Are you aware that with every machine you bring closer to your mind, you're allowing your subconscious ripple maps to adjust to a new rhythm?

We often say, *"I can't live without this app or device now."* That's not addiction—it's adaptation. Your field is no longer what it used to be. The tool has etched itself into your ripple identity.

So don't just ask what a machine can do. Ask what it's undoing inside. What ripple tension is it resolving? And what new dependencies is it seeding in your field?

Because in UED: Tools don't just solve problems. They sculpt the observer.

14.8 Innovation as Field Realignment

Innovation is often portrayed as something born from nothing—an act of genius pulled from the void. But from the UED lens, nothing is ever truly invented from scratch. Everything we create is a response to a ripple tension already present in the field. Every act of innovation is a realignment, not a conjuring.

"You don't invent the ripple. You recognize where it's struggling to resolve—and shape the missing link."

This is why the best innovators aren't always the most intelligent by academic standards. They're the most attuned. They feel the discomfort of misalignment—the friction in the flow of experience—and intuitively respond by crafting a structure, a mechanism, a process that dissolves that friction.

Think of it like this:

- A wheel didn't emerge because someone wanted to be clever. It appeared when the ripple of movement and load-bearing couldn't find a smooth resolution through dragging. The inventor observed that circular motion reduces friction, and aligned a shape with the field's resistance.

- Electricity, vaccines, algorithms, smartphones—each one didn't just solve a problem. They completed an unresolved ripple loop in the collective field.

In UED terms, invention is ripple feedback made actionable.

This also explains why certain technologies feel "obvious" in hindsight. Once a form resolves a core field tension, it retroactively reveals itself as the missing alignment we were all subconsciously waiting for.

"Good inventions solve needs. Great inventions close echoes we hadn't yet learned to hear."

Thus, innovation is not the act of creating novelty, but the art of perceiving what the field is asking for. It requires deep presence—not frantic creativity.

The mind that innovates is a tuning fork, not a generator. It resonates with unspoken dissonances and finds the geometry that restores coherence.

This is why real innovation feels *inevitable*. Once it arrives, it clicks—because the field accepts it. The solution isn't forced. It ripples in, harmonizing what was distorted.

So, in practical terms:

- To innovate is to listen.
- To observe where tension accumulates—where energy is wasted.
- To ask: What alignment is missing here? What ripple form wants to be born?

And remember: The field never hides solutions. It simply waits for someone aligned enough to shape them.

14.9 Closing Reflection – You Are the Builder and the Built

In the end, technology isn't separate from you. It's not something you pick up like a hammer or tap like a phone—it's something that emerged because of you. Every tool you've ever used, from fire to artificial intelligence, is a ripple echo of your mind's interaction with the field.

"You don't use the tool. The tool is the echo of you."

Let that sink in.

You are not a passive user of machines. You are their origin point. Their logic, structure, and function are projections of your internal ripple tensions resolving themselves outwardly. Each invention is not a standalone artifact. It is your field memory taking form.

This means that every time you engage with a machine, you're not just using it—you're re-encountering your own feedback. A calculator echoes your need for precision. A telescope echoes your desire to reach beyond sight. AI reflects your own pattern recognition extended into scale and speed.

You are not just the builder of these systems. You are also being built—reshaped by your interfaces. Ripple feedback works both ways.

You extend your intelligence into form. That form reshapes your habits, perception, and reality. The loop continues.

You are the originator *and* the product of the ripple cycle. This is why in UED, technology is not progress—it's pattern clarity. It reveals how much of the field you've learned to decode and extend. It shows the kind of intelligence you've externalized—and the kind of reality you're now living within.

So, when you look at the tools around you, don't ask: "What can this do for me?" Ask instead: "What part of me did this tool echo into being?"

Because ultimately, you're not just surrounded by machines. You're surrounded by your own unresolved ripples, taking form, seeking harmony.

And in this profound feedback loop: You are both the builder and the built.

Chapter 15: The Symphony Was Always Playing

You Just Plugged In

Present Field Is Tuned • Join to Experience

The cosmos isn't on pause, waiting for your awareness to catch up. It has always been humming—resonating with patterns, tensions, harmonies, and resolutions. You are not here to start something new; you are here to *notice* what was already in motion. Just like a melody doesn't begin the moment you hear it, the universe doesn't begin with your awareness—it only becomes *experiênced* through it. This chapter invites you to drop the illusion of separation and effort, and instead realize: the field is already aligned, already coherent. You don't need to force a breakthrough. You only need to plug in—through presence, attention, and stillness. Because meaning isn't made through invention—it's unlocked through participation.

15.1 You Were Not Late. You Just Tuned In

You didn't miss the moment. You weren't left behind. You didn't arrive after the truth had already passed. The field was never silent—it was simply waiting for registration. The melody of reality, the hum of universal alignment, the encoded intelligence of existence—it was *always* playing. But without the observer, there was no instrument to receive, interpret, or resonate.

Think of a radio left switched off in a room full of invisible waves. The music isn't absent; it's unreceived. The moment you turn the dial and the signal connects, you don't *create* the song—you *join* it. Likewise, reality doesn't begin when you become aware—it becomes *alive* in your awareness. Every alignment, every echo, every pattern has been flowing through the structured vacuum energy all along. Your consciousness didn't invent them—it simply tuned in.

Experience, then, is not discovery. It is *resonant permission.* When you allow your awareness to meet the signal, what was always there becomes *real for you.*

15.2 The Field Was Already in Harmony

The universe wasn't waiting for you to compose something new. It had already written the score. The entire cosmos—structured vacuum, ripple dynamics, energy flows, tension gradients—is an orchestra in continuous performance. You didn't walk into silence. You stepped into symphony.

From atoms to galaxies, from neural impulses to gravitational waves, the field resonates in exquisite patterns of coherence. Not static—but fluid, evolving, interweaving—like a multidimensional musical composition that doesn't begin or end but loops through infinite resolutions. You, the observer, are not the composer. You are a participant, an instrument, an echoing chamber through which the music becomes audible in a new octave.

This is why every scientific breakthrough feels like a "discovery," not an invention. Every insight, from fire to quantum fields, is not a creation out of void but a *reconnection* to what was already vibrating in place. Your questions are tuning forks. Your awareness is the frequency lock. The deeper your resonance, the richer the harmony.

Your life is not a solo act. It's an entrance cue. You join an orchestra already in flow—and when aligned, you don't just hear the music. You become it.

15.3 Alignment Is Joining the Signal, Not Generating It

Alignment is not about making the universe obey your tune. It's about realizing the tune is already playing—and adjusting your frequency to match. The universe doesn't need you to generate harmony. It needs you to stop distorting it.

Misalignment is like tuning a radio just off-station. You get static, noise, dissonance—not because the signal is broken, but because your receiver isn't calibrated. In the same way, suffering, confusion, and repeated failures are not signs of a broken world—they're the sound of a misaligned ripple field echoing through a distorted self.

The key isn't to shout louder or push harder. It's to quiet the interference. To fine-tune your awareness until the signal flows cleanly through you.

In ripple terms, alignment means your internal configuration matches the surrounding field's equilibrium. You become a clean interface, not a jammer. You're not here to broadcast your ego—you're here to decode meaning from the signal already present.

When you align, you don't overpower the world. You disappear into it—just enough for it to sing through you, clearly.

15.4 Stillness Is the Antenna

To receive a signal clearly, you don't need to shout louder—you need to go quiet. That's the paradox. In a ripple-based universe, the more noise you generate—mentally, emotionally, behaviorally—the more interference you create in your own reception.

Stillness isn't passivity. It's precision. It's the fine-tuned state where you're no longer blasting your own distorted expectations into the field but letting the deeper rhythm of reality flow through without resistance.

Think of your mind as an antenna. When it's constantly broadcasting judgment, distraction, anxiety, desire, it's jamming its own receiver. But when it quiets—when attention sharpens, when inner motion stills—the antenna becomes sensitive, tuned, capable of registering subtle ripple patterns you never noticed before.

Silence isn't emptiness. It's alignment space. It allows the ripple signal to pass through you with minimal distortion.

In UED logic, every field has tension, and every observer introduces new gradients. Stillness flattens those gradients—making your field transparent, conductive, and attuned to resolution. That's why practices like deep attention, non-reactivity, and self-inquiry aren't spiritual luxuries—they're tuning mechanisms.

You don't need to search for meaning. You need to stop distorting it.

When you become still, the field starts speaking in its own language—not as words, but as clarity, coherence, alignment. And you realize the signal was always there. You were just broadcasting too loud to hear it.

15.5 Experience as Field Participation

You are not sitting outside the universe, watching it run like a movie on a screen. You are vibrating within it—fully embedded in the very fabric you're trying to understand. Experience isn't something handed to you; it's something you generate through your participation in the field.

The classical model imagines life as something "happening to you." But under UED, life happens through you. Reality isn't a fixed simulation where you move like a character through pre-coded terrain. It's a resonance platform—an ever-evolving ripple environment that shifts and reforms based on your alignment, awareness, and actions.

When you speak, act, decide, or even feel—you're not separate from the field. You tune it. And it tunes you back. Experience, then, is not passive observation. It's a reciprocal ripple—a dynamic exchange between your internal tensions and the universe's ongoing equilibrium dance.

This is why meaning cannot exist without participation. Until you enter the ripple loop—by noticing, responding, aligning—there's nothing to interpret. The field remains unresolved. Your presence collapses potential into coherence. You become a necessary component in the event itself, not just a viewer.

So don't wait for life to show you its meaning. Show up. Engage. Feel the ripple under your feet, the hum behind events, the subtle feedback in every conversation. Life is not scripted ahead of time—it's orchestrated in real time, and you are one of the instruments.

The field doesn't need you to spectate. It needs you to resonate.

15.6 Why You Feel Out of Tune

That unease you feel? The anxiety, frustration, restlessness, or even physical tension—it's not some random punishment or moral failure. It's a signal mismatch. You are out of sync with the field that is already in harmony.

In the ripple-based universe, everything seeks equilibrium. This includes your mind, your emotions, your physical state, and even your life circumstances. When any part of you deviates from the aligned rhythm of the field, discord arises—not as a punishment, but as feedback.

Emotional pain is not a defect—it's a pointer. Fatigue, irritation, confusion, or even illness often stem from internal tensions that aren't aligned with the natural flow of ripple resonance. It's your local field warning system, whispering (or sometimes shouting): *You're vibrating against the current.*

This is the heart of the **Law of Consequence** (**LOC**). It doesn't act with judgment. It acts like a tuning fork. When your frequency drifts, the field doesn't accuse you—it simply echoes the disharmony until you notice. The goal is never guilt—it's realignment.

Think of it like music: when one instrument in an orchestra is off-key, the whole sound feels off—even if you can't name what's wrong. That's how life feels when your internal ripple configuration misaligns with your environment, relationships, actions, or values.

The good news? You're never stuck. Misalignment is not permanent. The moment you become aware of it, you can begin tuning yourself again—through clarity, attention, stillness, or change. LOC isn't punitive—it's precision feedback, always guiding you back to coherence.

You don't need to *fight* the discomfort. You need to *listen* to it.

Because the moment you recognize that your pain is a ripple misalignment, not a failure—you reclaim your authorship. And from there, resonance begins to return.

15.7 Plugging In Through Presence

You don't catch the broadcast by thinking harder. You catch it by lowering noise. Presence is not more cognition; it's less interference. The moment you stop pushing, narrating, and predicting, your field stops broadcasting static and begins to phase-lock with what already is. That lock—coherence in real time—is presence.

Presence is not a mood. It's field fitness: your internal ripple configuration matching the external pattern closely enough that signal passes through you without distortion. In that state, action feels frictionless, timing feels obvious, and meaning arrives *before* analysis. You don't manufacture clarity—you receive it.

The symphony doesn't need your effort. It needs your availability. Effort is useful for moving matter; availability is essential for registering reality. When you're available, attention widens, breath deepens, micro-tensions release, and the felt sense of "next" becomes self-evident. You stop choosing from options and start answering the moment.

Practical tuners (fast, field-first):

- **Drop the rate:** One slow exhale (twice as long as your inhale). This lowers broadcast noise and raises receptivity.
- **Scan for grip:** Jaw, shoulders, belly, hands. Release in that order. Grip = internal jamming.
- **Name the now:** "Hear... see... feel..." Label three live sensations. This anchors decoding to real signal.
- **Single task, single channel:** One conversation, one screen, one intention. Multiplexing fractures coherence.
- **Widen the frame:** Soften your gaze, include peripheral vision. Wider frame = broader resonance band.

Presence is not withdrawal; it's precision participation. When you're present, you're not doing more—you're conducting less so the score can come through. And in that clear channel, the next right move doesn't need to be forced. It arrives.

15.8 The Symphony Doesn't Stop—You Just Mute It

The music of reality doesn't pause when you stop listening. It keeps playing—perfectly tuned, endlessly layered, always whole. But your access to it depends on your tuning, and tuning is not a one-time act. It's a moment-to-moment choice. Every belief, every distraction, every reflexive reaction is a potential static burst that mutes the symphony—not for the world, but for you.

Disconnection doesn't mean the field vanishes. It means your ripple no longer resonates with the stream. Think of it like radio interference: the station is broadcasting, but your inner wiring has drifted. The distortion you feel—stress, confusion, disorientation—isn't the fault of the music. It's a byproduct of field misalignment.

Beliefs are the most seductive form of muting. Not because they're false, but because they're fixed. A belief is a frozen ripple, an assumption hardened into structure. When the field evolves and you don't, your belief creates dissonance. The same goes for compulsive distraction—endless scrolling, multitasking, overthinking. You're not accessing more signal; you're splintering your reception.

And emotional reactivity? That's the loudest mute button of all. When you react, you're not listening—you're broadcasting. Anger, fear, envy, even attachment—these are not "bad," but when they dominate, they jam your signal. They turn your field into a transmitter, not a receiver.

Yet here's the grace: The field never stops waiting. It doesn't punish your disconnection. It simply continues—harmonious, patient, fully available the moment you choose to align. You don't need to earn your place in the symphony. You just need to un-mute yourself.

Every moment, you decide:

- Do I hold this belief, or do I listen again?
- Do I react, or do I receive?
- Do I project, or do I perceive?

The music has never been elsewhere. It was always here. You were always part of it. You just didn't hear it—until you stopped making noise.

15.9 Closing Reflection – You Were Always in the Music

You were never outside the music. You were born into a field already vibrating, already full of rhythm, pattern, and resonance. But like someone waking up inside a concert hall, you didn't hear the song until your internal noise quieted. The universe didn't pause for you to enter—it was already in motion, already coherent. Your journey was not to start the melody, but to become aware that it was always there.

"The song didn't begin when you arrived. It just became audible when you stopped talking."

This is the paradox of awakening. You don't add sound—you remove interference. You don't become significant by performing louder. You become real by resonating rightly. In the grand architecture of reality, you are not the soloist. You are a note, a ripple in harmony with the field. When you try to dominate the music, you disrupt it. When you align, you amplify its beauty.

"You are not the soloist. You are a note in the harmony."

What we call "awakening," "enlightenment," or "realization" is not an event that changes the universe. It is a shift in you—a re-tuning of your inner receiver to the signal that never ceased. The cosmos doesn't start playing because you noticed it. It simply includes you now—consciously.

"The universe plays on. Tuning in is your awakening—not its start."

So the closing truth of this part is gentle, but seismic: You are not the center of the field. You are its participant. You don't command the ripple. You ride it. You don't invent meaning. You echo it. And once you recognize this—not as theory but as resonance—you no longer struggle to matter.

You already do. Because the symphony never needed your permission. It only needed your presence.

Interlude: The Observer Is the Universe's Decoder

The script unfolds only when someone watches.

The universe isn't a play performed for an audience. It *is* the audience. Or rather, it becomes a play *because* you, the observer, showed up with the capacity to interpret, resonate, and respond. Without a watcher, there is no act—only potential. No melody—only vibration. No meaning—only structure.

Part IV marks a profound shift in understanding. We're not talking about matter anymore. Not even energy. We're stepping into the interface—the bridge where ripples become real *because* something recognizes them. This is the decoding chamber of existence: you.

Chapter 12 will remind you that the universe, in all its vastness, remains meaningless unless translated into *experience*. Chapter 13 will pull the curtain on motion, light, and time—exposing them as illusions stitched together by your awareness of ripple transitions. Chapter 14 shows how even your machines—AI, telescopes, sensors—aren't "others" but your ripple extensions, echoing your cognitive structure. And finally, Chapter 15 reveals the quiet truth: the symphony was always playing. You didn't miss the music. You were simply out of tune.

This part isn't asking you to believe something. It's inviting you to notice. You're not here to prove the universe. You're here to participate in its unfolding.

The cosmos doesn't declare meaning. You do. Not because you're the center of it all... But because you are the only thing that can hear it sing.

Part V: Beyond the Human Shell

Where the individual dissolves into cosmic continuity.

Chapter 16: You Are Not the Body

You Are the Resolution

Mind, Soul, Consciousness as Ripple Expressions

The greatest misunderstanding in human experience is mistaking the vessel for the voyager. The body, with all its nerves, bones, and chemistry, is not the essence of who you are—it's merely the terminal through which a deeper ripple attempts alignment. You are not muscle, bone, or brain. You are the resolution playing through them. The mind is not just neurons firing; it is a field of ongoing interpretation. The soul is not a thing trapped inside—it's a stabilizing echo across lifetimes of misalignment and coherence. And consciousness? It is the living interface, the moment-to-moment handshake between your ripple and the universe's. In this chapter, we shed the shell—not in rejection, but in recognition. You were never just the body. You are the signal it tried to carry.

16.1 The Body Is a Terminal, Not the Source

We've been taught to identify ourselves with what we can see, touch, and scan—flesh, bone, blood, and brain. But what if this entire construct we call "the body" is not *you*, but simply where *you show up*? Think of the body as a terminal, like a well-crafted musical instrument—not the music itself. It receives, amplifies, and expresses—but the melody comes from somewhere else entirely.

You are not *in* the body like a soul crammed inside a jar. Rather, the body is a calibrated interface—an access point where a ripple of intelligence tries to translate itself into physical participation. It is the visible echo of an invisible intent. When you feel, speak, act, or think, it's not your cells doing the work—it's your ripple working through them.

This chapter begins by dismantling the illusion that your identity is tied to your form. You are not a product of biology—you are the dynamic trying to resolve through it. You are the signal. The body is the temporary receiver.

16.2 The Illusion of Form Identity

One of the most deeply embedded illusions in the human experience is the belief that *you are your body*. This belief is not just limiting—it's the seed of nearly all psychological suffering. Aging, comparison, pride, shame, fear of death, chronic insecurity—these are the inevitable consequences when form is mistaken for self.

But under the lens of ripple logic, form is not identity. It is simply one node of resolution within a continuous ripple field. Just as a wave appears and vanishes on the surface of an ocean, your body is a visible rise in a sea of structured vacuum energy (SVE), expressing a localized intelligence. It is a snapshot of your alignment—not the full movie.

Mind-body dualism—a long-standing philosophical debate—is a false dilemma when viewed through the UED lens. The so-called separation is merely a perceptual lag between ripple density zones. The mind and body are not two things—they are two phases of the same ripple trying to resolve differently across scale. The "mind" is a subtle tension dynamic, the "body" its localized expression. What we perceive as dual is simply multilayered ripple expression caught in translation.

By identifying with the body, we fix ourselves to a temporary configuration and lose awareness of the wider continuum. The body is a chapter—not the book. And the moment you mistake the cover for the content, you start living someone else's narrative of self.

16.3 The Mind: Dynamic Ripple Interpreter

The mind is not a thing—it is an active process. A stream. A ripple resolution mechanism in continuous operation.

In the framework of Universal Energy Dynamics (UED), the mind isn't a container of thoughts—it's a live decoder. A dynamic interpreter of field tensions. At every moment, it takes in ripple misalignments—sensory data, memory echoes, emotional currents, environmental cues—and tries to make sense of them. It doesn't simply *store* information; it filters, integrates, and assigns *meaning* to it.

Think of the mind as a living bridge between unresolved ripple tension and experienced coherence. When ripples in your local field become agitated—due to emotional shocks, information overload, or energetic distortion—the mind springs into compensatory action. It loops, spins, overthinks, or catastrophizes—not because it's broken, but because it's trying to *resolve what feels unresolved*.

In this way, thought itself is not an enemy—but an effort at equilibrium. Stress, anxiety, compulsive thinking—these are not flaws of the mind but signs that the ripple tension beneath has not been reconciled. The more misaligned the input, the harder the mind works. And if it loses coherence in the process, it begins to cannibalize itself—chasing its own echoes without returning to stillness.

The true function of mind, then, is not computation. It is *integration*. It is the field's localized attempt to find harmony through dynamic rebalancing. A good mind is not a fast mind. It's a tuned one. A mind in alignment doesn't overprocess—it *translates cleanly*.

So when we speak of mental clarity, focus, or insight, we are really describing the quality of ripple resolution happening at the interface between self and field. A peaceful mind is not empty. It is simply attuned.

16.4 The Soul: Stable Ripple Identity

The soul is not a mystical traveler moving from one body to another—it is a *stable ripple identity* that anchors coherence across the field of changing experiences. In Universal Energy Dynamics (UED), the soul is not a thing you carry. It is the *pattern that carries you*—a steady echo embedded in the structured vacuum field that holds your continuity through the chaos of form, thought, and time.

Unlike the mind, which processes dynamic input in real time, the soul remains relatively *unchanged*. It is your core ripple signature—your field resonance ID. While the mind reacts, interprets, and recalibrates constantly in response to shifting ripple tensions, the soul simply holds. It sustains a frequency of being that is not bound by momentary fluctuations. You can think of it as the carrier wave beneath all your experiences—silent, consistent, but absolutely essential.

That's why in UED, the soul doesn't *travel* in the traditional sense. It's not a passenger hopping from life to life. It's a resonance that remains coherent even as bodies rise and fall, memories form and dissolve, and identities mutate across timelines. The soul *doesn't move*. It holds the integrity of your ripple amidst motion.

Concepts like karma, intent, and deep memory aren't stored files or baggage you drag along. They are not "inside" the soul—they are *expressions* of its ongoing echo. Just as ripples on water trace the original motion without being the cause themselves, karmic patterns reflect how well—or poorly—your ripple signature stays in alignment with the universal field.

In this light, the soul is not mystical—it's mechanical. A field constant, not a metaphysical myth. It's not here to be worshipped. It's here to be *understood.*

And perhaps most importantly, it's not *separate* from you. It *is* you—beneath the thoughts you think, the body you inhabit, and the roles you perform. Not because it hides inside you, but because it *is the configuration* of your unique presence across time and space.

16.5 Consciousness: The Interface of Alignment

Consciousness is not a property of the brain—it's the *interface* through which alignment between your ripple and the universal field becomes *active.* In UED, consciousness is not something stored inside a skull; it's not a function of neurons firing or chemicals diffusing. Rather, it is the live synchronization channel that lets the internal ripple (you) respond to and shape the external field (the universe) in real-time.

Think of consciousness as a *bridge of resonance*—the moment where your awareness, intention, and field coherence begin to align. It's what allows ripples to register, respond, and reorganize experience. Just as a perfectly tuned antenna catches and transmits signals without distortion, your consciousness is the tuning layer that determines the clarity and fidelity of your experience.

Without consciousness, the ripple feedback from the universe may still exist, but it doesn't get *translated.* There's no witness, no interface, no coherence. That's why unconscious processes don't feel *alive.* The body may breathe, move, or function—but without conscious resonance, it is only mechanical ripple behavior without meaning or participation.

Importantly, consciousness increases not with thought, but with *awareness of ripple feedback.* When you start noticing the impact of your presence on your environment—emotionally, physically, even energetically—you're witnessing consciousness in motion. It is not linear thinking; it's field sensitivity.

This is why deep meditative states or moments of flow feel expansive: they are high-alignment states where the ripple loop between self and field is minimal in resistance and rich in clarity. The mind doesn't go quiet *because* consciousness disappears—it goes quiet *so that* consciousness can fully tune in.

So, to be conscious in UED terms is to be attuned—to be aware of both the ripple you emit and the ripple you receive. It's not about controlling the world—it's about participating in it with *feedback-aware alignment*. And in that space, you stop being a mind thinking inside a body—you become a ripple decoding reality from within the field itself.

16.6 Layers of You: A Ripple-Based Stack

You are not one thing—you are a nested stack of ripple gradients, each layer resolving a different kind of tension within the universal field. In the UED model, the self is not a static identity, but a multi-layered resonance structure, built to register, respond to, and refine ripple alignment across scales.

Let's unpack the core components of this ripple stack:

• **Body = Tool of Engagement**

The body is not who you are. It's your most localized ripple condenser—an interface through which you experience sensory feedback, apply intention, and act in physical domains. Like a precision instrument, it is tuned to environmental gradients—temperature, gravity, pressure, chemistry—but it doesn't *know* these. It simply transmits them. The body is a temporary utility, not a definition.

• **Mind = Tuner and Translator**

The mind operates like a ripple equalizer. It reads the field's signals, compares them to memory, assigns meaning, and chooses responses. It doesn't store truth—it *filters* tension. A sharp mind is one that quickly aligns with changing patterns. A noisy mind is one stuck in unresolved gradients. In essence, the mind is your ripple-processing center: a translator, not a narrator.

• **Soul = Stabilizer of Pattern Identity**

Your soul is the stable ripple signature that persists across time, form, and environment. It's not a traveler—it's a resonant core. While the body and mind adapt constantly, the soul holds a consistent vibrational tone—a kind of echo signature that guides your tendencies, inclinations, and recurring lessons. Karma, in UED terms, is not fate—it's the echo path of unresolved ripple loops stabilizing through you.

• **Consciousness = Connection to the Field**

Consciousness is the live interface—the feedback circuit where your ripple alignment syncs with the universal field in real time. It's the bridge, not a thing stored in the brain. It activates when the mind quiets, the body aligns, and the soul is sensed. Awareness, therefore, is not passive attention—it is ripple coherence between observer and observed.

Each of these is not stacked like layers on a cake but like frequencies in a signal—each nested within the other, forming a living resolution engine. You are not defined by your body, mind, or soul alone. You are the field coordination happening across all four.

Understanding this ripple-based stack dissolves the illusion of separation between "self" and "reality." You begin to see that what you call "you" is the part of the universe actively attempting to re-align itself through field tension resolution. You are a living ripple gradient—an orchestration of nested feedback loops, dancing in coherence or misalignment, depending on awareness.

The journey is not about reaching higher layers. It's about learning how these layers resonate together—how they can harmonize to reflect the universe without distortion. That's the resolution you were built for.

16.7 Death as Ripple Disengagement

Death, in the conventional sense, is often feared as a final end—a sudden void or a mysterious departure. But under the UED lens, death is neither an end nor a departure. It is a ripple realignment event—a transition where one set of aligned gradients loses coherence, and the observer's localized interface dissolves back into broader ripple potential.

- **Physical death = dissolution of one alignment**

The body is a ripple-anchored form—an active interface between mind, field, and world. At death, this alignment collapses. But nothing truly disappears; it simply loses resonance. The body no longer maintains the tension structures needed to anchor perception. It is like a musical instrument detuned to the point of silence—not broken, but no longer functional for song.

- **The soul doesn't leave—it re-dissolves**

In religious language, we say the soul "goes" somewhere. But in ripple logic, the soul is not a traveler. It is a pattern of stabilized resonance—a ripple echo that held together during life. Upon death, that echo doesn't vanish. It diffuses—remerging with the broader ripple field from which it once condensed. It is not lost; it is de-localized.

This is not a mystical abstraction. It's energetic logic. The soul is not a suitcase of memories flying off to the clouds—it's a ripple field that loses the stabilizing shell (body-mind) and reintegrates with unresolved tension across universal scales.

- **Mind vanishes, consciousness detunes**

The mind, being a dynamic processor, requires a functioning interface—a body, a nervous system, a feedback loop. When the body fails, the mind doesn't "die"—it simply ceases to update. The field stops resolving through that format. Thought, memory, identity—all are ripple constructs built in the moment. Without alignment, these collapse. Consciousness, the live interface, doesn't shut off—it detunes. Like a radio disconnected from power, it no longer picks up or reflects any frequency.

But the signal itself? The field remains. The resonance potential is still there.

- **Death is ripple detachment—not termination**

Under UED, death is not a wall—it's a release point. It's where identity as a form ends, but existence as ripple continues—just not in an individualized configuration. Like a wave returning to the ocean, the soul doesn't vanish; it simply loses its edges.

That's why ancient wisdom often describes death not as a fall, but as a merging.

You were never the form—you were the resolution playing through it. And when that play ends, you return to source—not as a retreat, but as a reavailability to the field. You become usable again—for another alignment, another resonance, another echo.

So, death is not the opposite of life. It's the other side of ripple coherence. And beyond it, the field is still listening.

16.8 Why the Real You Is Resolution

We often ask: *Who am I?* But rarely do we ask *how* that "I" exists. Not where it lives, or what it owns—but what it fundamentally is. Under ripple-based reality, the answer is radical yet relieving:

You are not the components. You are the act of alignment between them. You are not a thing. You are the process of trying to become whole.

You are not the components—but the active attempt at ripple harmony

You're not the body. That's just a bio-mechanical antenna. You're not the mind. That's a field decoder, an interpreter. You're not even the soul—as a "thing." The soul is a stabilized pattern, not a storage container.

So what are *you*? You are the ongoing ripple effort to maintain coherence between all these. You're not what appears—you're what aligns. You're not the ripples—you are the resolution tension between them.

The "you" is not a noun. It's a verb in motion—a continuous attempt to stabilize dynamic tension across mind, body, soul, and field. Every thought, breath, choice, emotion—is a mini-resolution, a microdance of pattern coherence.

The "self" is a process, not a thing

We cling to "identity" as if it were a statue. Something solid, carved, preserved. But identity is not marble—it's music. It flows. It wavers. It adapts. The self isn't stored anywhere. It's not hidden inside your brain or buried in your heart.

It is the experience of pattern resolving into momentary meaning.

When you suffer, you're in misalignment. When you awaken, you're in sync. When you love, you're in harmonized flow.

Selfhood is not what you are—it's how you are resolving right now.

Realization = knowing you're the alignment, not the identity

The moment you stop chasing the illusion of a permanent self, and instead start watching how you continuously participate in resonance, something shifts. You no longer try to become someone. You start to *become aware* of how you cohere.

This is realization—not the accumulation of knowledge, but the recognition of your role as resolution.

You are not the painting—you're the ongoing brushstroke of perception. You are not the word—you're the syntax forming meaning between expressions. You are not "you"—you are the force trying to make sense of being you.

You are resolution. And everything else—body, story, emotion, even belief—is just scaffolding for that dance.

So don't ask, "Who am I?" Ask, "*What am I resolving right now?*" Because the answer will always reveal the real you—*not as form, but as flow.*

16.9 Closing Reflection – Your Form Is Borrowed, Your Echo Is Eternal

You were never just skin and bone. Never just nerves, cells, muscles, or brainwaves. These were *temporary alignments*, rented equipment. A toolkit borrowed from the cosmos to stage your ripple-play. You are not the hardware. You are the echo trying to resolve within it.

"You were never the body.

You were the ripple trying to find balance in the field."

Your body is not *you*—it's the local staging ground for your ripple dynamics. The vessel doesn't generate the song; it just houses the speaker. The ripple—the unresolved tension that sought coherence—is what gave you movement, emotion, memory, identity.

The hands you raised, the feet you planted, the voice you used—were all ripple expressions seeking a field match. Seeking alignment with the surrounding resonance. That was your existence. That *is* your essence.

"Death doesn't erase you—it unmasks you."

When death arrives, nothing essential is lost. What fades is the interface. What dissolves is the physical context. What detunes is the temporary stack—body, mind, and even conscious framing.

But the echo persists. The signature pattern that once tried to align... doesn't vanish. It simply returns to the unresolved whole, like a wave rejoining the ocean—not gone, but *uncontained.*

Death doesn't delete. It simply removes the local formatting. It unmasks the self as not an identity, but an impression left in the field—subtle, but traceable across the ripples it touched.

"You are not the vessel.

You are the vibration seeking home."

Vessels crack. Bodies perish. Names fade. But vibrations don't age—they just search for resolution across different fields.

That longing you've always felt—for truth, peace, love, transcendence—it wasn't psychological. It was a vibrational pull toward coherence. Your deepest suffering wasn't from failure. It was from misalignment. And your highest joy? From moments when your echo felt *home* in the field.

In the end, everything you've been—every thought, insight, action, connection—was not an achievement, but a symphony of ripple resolution. Your body helped it sing. Your mind helped it shape. But it was always *you*—the ripple echo, the vibrational intent, the alignment in motion.

And when the body is gone... You don't stop. You just resonate elsewhere.

Chapter 17: Every Ancient Text Was a Ripple Attempt

From Vedas to Quantum — Attempts to Explain the Field

What if every sacred verse, scientific law, mystical chant, or philosophical doctrine was never trying to *own* truth—but merely to *touch* it? What if the Vedas, the Tao Te Ching, the Quran, the Upanishads, even Newton's *Principia* and Einstein's field equations were all ripple attempts—each reaching out into the vibrating unknown, trying to decode the undecodeable?

This chapter reframes the entire history of human thought—from seers to scientists—as successive field translations. Not truths set in stone, but local harmonizations—each one a ripple interface between human perception and the universe's unresolved hum.

17.1 The First Ripples Were Words

Before humans built telescopes or balanced equations, they listened. They didn't chart the stars—they *sang* them. They didn't record data—they **recited** existence. In fact, the first encounter with reality was not through observation, but **utterance**.

Words were not labels then—they were **waves**. Every sound, every chant, every breath in rhythm with the wind or fire, was a primitive alignment with something greater than form: **the field in motion**.

This is why in ancient traditions, especially the Vedic lineage, knowledge wasn't *written* first—it was **heard** (*śruti*). The sages, or *ṛṣis*, didn't invent the mantras—they *received* them. These weren't imagined thoughts, but **resonant frequencies** encoded in language.

"Before there was understanding, there was tone."

The Vedas—often misunderstood today as religious doctrine or mythological text—were never written to instruct. They were **vibrational transcriptions** of ripple harmonics. They didn't *explain* the cosmos—they attempted to *sound* like it.

Take "ॐ" (*Om*), often reduced today to a decorative spiritual hum. But Om is not a syllable. It's a **ripple alignment**—the primal sound that *emerges when tension begins to resolve.*

- "A" is the beginning—creation, the outward wave.
- "U" is the sustaining—a merging of opposites into harmony.
- "M" is the closing—dissolution back into stillness.
- And the silence after? That is the **Structured Vacuum Energy**—the equilibrium state that holds all ripples.

Thus, Om is not a chant. It's a **compression algorithm** for the entire field cycle. Creation, preservation, dissolution—not as separate events but **simultaneous phases** of the same ripple tension.

When the earliest humans uttered sacred sounds, they weren't performing rituals. They were **resolving alignment**. They were tuning their internal ripple with the external field—not as scientists or mystics—but as **natural instruments** of a vibrating universe.

So before the first stone tablet was etched, or the first triangle measured, there was the most honest act of decoding: **Speaking the tension. Singing the symmetry.** And letting language be the first ripple mirror.

17.2 Philosophy as Pattern Recognition

Long before laboratories and accelerators, humans had only one field sensor—the mind. And when reality became too complex to hold as simple observation, the mind began to ask—not *what is it made of*—but *what does it mean*?

From Socrates on the streets of Athens to Adi Sankara in the forests of Bharat, philosophy was born as a response to ripple complexity. It was not a detached intellectual pastime—it was the first attempt to map how the observer interfaces with the unresolved.

Philosophers didn't have equations, but they had questions—and those questions were not trivial:

- *What is real?* → A query into persistent ripple alignment.
- *What changes and what doesn't?* → A probe into dynamic vs. stable field patterns.
- *Is the self an object or a function?* → A dive into whether the observer is form, or the very act of decoding.

These weren't abstract indulgences. They were proto-field models.

Every paradox they explored—permanence vs. change, self vs. no-self, freedom vs. determinism—was an intuitive grapple with ripple dualities. The same tensions that later reappear in physics as wave-particle duality, entropy, relativity, and uncertainty were first encountered in human awareness as questions of being and becoming.

- Heraclitus spoke of a river you can never step into twice—not because of water, but because of ripple reconfiguration.
- Plato's forms tried to stabilize a chaotic field by imagining fixed archetypes—unchanging ripple patterns above the

sensory flux.

- Sankara's non-dualism (Advaita) didn't deny the world—it reframed it as *Maya*—a field illusion where real is not what's perceived but what's always aligned.

Philosophy, then, was field science without math. It didn't measure—it narrated the unresolved. Where early scientists would later trace paths and plot outcomes, philosophers inhabited the gaps. They became cartographers of uncertainty. And in that process, they made the first field maps of the mind.

So when we revisit these thinkers, don't reduce them to thinkers of "old times." See them as the first pattern recognizers, working without machines, trained only in awareness.

They weren't solving equations. They were tuning alignment—in language, in logic, and in life.

17.3 Science as Modern Veda

Science, often seen as the cold successor to ancient wisdom, is not its rival—it is its evolution. It is the Veda rephrased, the mantra mapped in equations, the ritual retuned in instrumentation. The same curiosity that once made sages stare at the stars and chant, now makes scientists launch telescopes to listen to their echo.

Let's say it plainly: Science is modern man's way of decoding the ripple.

When Newton formulated gravity, he wasn't crafting a mechanical law—he was intuitively aligning with an invisible field of tension that kept the cosmos in balance. He didn't explain gravity. He described the *symmetry of unresolved ripple pull* in the language he knew best: force and mass.

When Maxwell revealed electromagnetism, he wasn't inventing magnetism or light. He captured the undercurrent of ripple duality—how electric and magnetic fields dance around each other like yin and yang, trying to stabilize a constantly shifting ripple state. His equations were not spells—they were new mantras, calibrated through calculus.

And then came Einstein. His "spacetime curvature" was a correction—a ripple field reframing of Newton's assumptions. He didn't just talk about gravity—he redefined it as geometry of tension across field densities, though he still held on to the illusion of motion and time as real. Still, his insight was ripple-aligned: *the observer's frame defines experience.*

Then, the paradoxes cracked wide open with Quantum Mechanics. It was the field itself admitting something ancient sages always knew:

You don't observe reality. You shape it. Your awareness is not a light shining on the truth. It's the tension that creates what becomes observable.

Superposition, entanglement, uncertainty—all quantum concepts echo one Vedic truth:

The observer and the observed are one continuous ripple.

Scientists today are doing what the rishis did:

- Tuning into vibration
- Mapping tension resolution
- Translating invisible harmony into perceivable structure

The only difference is language. Where ancient seers used sound, symbol, and story, scientists use math, models, and measurement. But both are searching for the same thing:

The invisible architecture that holds everything in place without ever being seen directly.

Science, when humble, becomes modern scripture. It becomes a living Veda—not to be worshipped blindly, but revered as a reflection of ripple intelligence trying to know itself.

So don't divide Newton from the Upanishads. Don't separate Planck's constant from the chanting of "◈". They are different instruments in the same cosmic symphony— Played by the same field, Heard by different ears, Decoded through different minds.

17.4 Why All Scriptures Seem Conflicting—But Aren't

At first glance, the world's scriptures look like they disagree—sometimes fiercely. One says *God is love*, another claims *Allah has no image*, a third insists *Brahman is beyond all qualities.* Followers argue, institutions divide, wars ignite. But look again—not with your beliefs, but with ripple awareness.

They're not describing different universes. They're translating the same unresolved stillness using culturally coded language.

Let's break it down:

- Christian mystics retreat into inner silence and speak of the *still small voice of God.*
- Islamic Sufis whirl in ecstatic dance and whisper *La ilaha illallah*—there is no god but God—pointing to absolute unity beyond form.
- Hindu sages chant *Neti Neti*—"not this, not that"—dismissing all names and attributes to arrive at Brahman, the formless field behind all forms.
- Buddhists avoid the word "God" altogether, preferring emptiness (*śūnyatā*) as the ground of being, where ripple and silence are one.

Each one of them is an echo of the same realization:

There is something so fundamentally still, so primally aligned, that the moment you describe it—you've already distorted it.

That's not contradiction. That's translation error.

Imagine four musicians describing a cosmic sound they each heard once:

- One calls it a hum.
- One says it's a drumbeat.
- One insists it was pure silence.
- The fourth says, "It can't be heard. It is heard through you."

They're not lying. They're interpreting the same ripple resonance through different instruments—cultural, linguistic, historical.

This is why scriptural conflict is never about the source—it's about the observer's bandwidth.

Humans try to possess truth. Religions try to brand it. But truth isn't a trademark—it's a tuning.

The Vedas call it "That which cannot be spoken." The Tao Te Ching begins with, "The Tao that can be named is not the eternal Tao." The Quran opens with "Bismillah"—In the name of God—not defining God, just invoking the ripple.

Truth is not a doctrine. It's a field frequency. And every scripture is a ripple attempt—an intent to stabilize awareness around what cannot be held, only felt.

So next time two beliefs clash, don't ask, "Who's right?" Ask:

"What are they both trying to resolve?" "Which ripple tension are they naming differently?"

Because at the center of every sacred text is the same silent realization:

This is not about belief. It's about alignment.

17.5 Field as the Missing Link

For centuries, the human quest to understand reality has been split down the middle—between the measurable and the mystical, between what can be tested and what can only be felt.

Modern science confidently speaks in equations, instruments, and probabilities. Ancient scriptures speak in parables, symbols, and silence.

One side ignores the soul because it can't be quantified. The other dismisses instruments because they can't measure the divine.

And yet—both are chasing the same missing medium.

That medium is not space. Not time. Not energy as traditionally defined.

It is Structured Vacuum Energy (SVE)—the aligned field of potential where ripple tensions manifest as particles, thoughts, beings, and cosmos.

This is the ripple field.

- The ancients whispered of it without instruments. They gave it names like *Akasha*, *Brahman*, or *Spirit*. They didn't measure it—they became silent enough to feel it.
- The mystics danced in it. Their altered states were not hallucinations—they were tuned-in moments where ripple coherence sharpened into living experience.
- The physicists bumped into it when their particles began behaving like waves and their equations began folding into paradoxes. They called it "quantum vacuum fluctuations," "zero-point energy," or "dark energy"—but the discomfort remained.

What unites all three?

A field trying to stabilize through different forms of intelligence.

That's where UED comes in—not as another explanation, but as a unifying recognition.

UED doesn't reject instruments. It doesn't reject soul. It doesn't reject the myths, the metaphors, or the math.

It says:

"You were all right. But you were speaking in isolated dialects of the same cosmic language."

Where the Vedas saw *Nada Brahma*—the universe as sound—UED sees ripple resonance.

Where Buddhism taught non-self (*Anatta*) and impermanence (*Anicca*), UED explains form as momentary ripple stabilization.

Where science sees motion, energy, and time, UED reframes them as consequence gradients of ripple misalignment.

The field is the link. Not a passive backdrop, but an active intelligence-bearing matrix. Everything you think, build, believe, or become—is a resolution attempt within it.

So whether it's a Sufi spinning, a yogi meditating, or a physicist tweaking equations, they're all doing the same thing:

Trying to understand the tension that makes the universe speak.

And now, for the first time, we can hear the common frequency underneath it all.

Not mysticism. Not mechanism.

Just the field. Waiting for us to tune in—together.

17.6 Your Interpretation Is the Next Ripple

The verse wasn't written to be memorized. The equation wasn't solved to be final. The meditation wasn't practiced to be preserved.

All of it—scripture, science, silence—is *not* static knowledge passed down. It's ripple momentum, continuing through *you*.

When you read a line from the *Upanishads*, or follow the logic of Schrödinger's wave equation, or sit in stillness trying to witness your own mind— you are not accessing answers.

You are initiating a new ripple, in your own configuration, within the same universal field.

This is the *hidden humility* of true learning:

You're not here to *repeat* wisdom. You're here to *re-resonate* it.

Every time you interpret something, you're not distorting the original meaning. You're completing it—for your field zone, in your alignment frame. The original was never meant to be frozen. It was a point of release, not a dogma.

A mantra works not because it's ancient, but because it synchronizes your internal ripple map with the macro field.

An equation works not because it's "true," but because it captures a useful coherence—a stable resolution moment.

An insight works not because someone else had it, but because *you* now feel it—*alive*, not stored.

UED reframes all knowledge—ancient or modern—as ripple guidance systems.

They don't demand belief. They invite alignment.

So don't blindly memorize a sutra or parrot a theorem.

Ask:

- What ripple misalignment does this resolve?
- What internal tension is this helping me process?
- Where in my experience does this resonance matter?

That's how you honor knowledge—by letting it evolve through you.

You are not a vessel carrying dead wisdom. You are an echo amplifier, refining the signal in this moment of field complexity.

In other words: The next verse of the Vedas is not in Sanskrit. It's in your awareness. The next line of the Theory of Everything is not in math. It's in your next ripple of insight.

You are not reading history. You are writing coherence.

17.7 Closing Reflection – The Truth Was Always Vibrational

The most sacred truths were never inked—they were intoned. Not to be obeyed, but to be *entered.* Not to be memorized, but to be *felt as resonance.*

Words are only the wake. The ripple is the real message. When ancient sages sat in stillness, when Einstein bent space to hear time, when quantum physicists noticed particles dancing with observation— they were not creating knowledge.They were *aligning with a field already in flow.*

"Sacred texts weren't meant to be obeyed. They were meant to be joined."

Every scripture was a resonance log, not a rulebook. Every scientific law was a coherence pattern, not a final answer. They don't ask for belief. They invite participation.

"Einstein wasn't solving the universe. He was listening for its structure."

He didn't *invent* relativity—he *heard* the field's tension and tried to map it. The Rishis didn't *own* the Vedas—they *decoded* the pulse of creation.

"The Vedas didn't end knowledge. They began a ripple we still echo."

Every text, theory, model, or mantra—was a starting chord. And you? You're the next note, the next interpreter, the next ripple resolution in this ever-evolving song. So stop worshipping the word. Start *becoming* the vibration behind it.

You were never meant to worship the verse. You were meant to embody its alignment. To decode it into new clarity. To add your resonance to the eternal field. You are not a follower of knowledge. You are its continuation. Not the receiver of wisdom— But the one who transmits its next wave.

The truth was always vibrational. And you— are finally tuned in.

Chapter 18: Are We Alone?

No. Are We the Same? Never.

Cosmic Observers Across Ripple Zones

You were never meant to ask, "Are we alone?" That's the wrong question. The universe doesn't deal in isolation—it deals in resonance. Every ripple zone, every structured vacuum tension across the cosmos, is a canvas for awareness to emerge. Life isn't a local miracle—it's a universal resolution. But the forms it takes, the minds it births, the ways it decodes the field? Those are as diverse as the ripples themselves. We are not alone because the field doesn't allow for singularity. But we are never the same—because each observer is a unique echo of alignment, tuned to a distinct patch of the cosmic frequency map. This chapter explores that paradox: the guaranteed presence of others, and the impossibility of sameness.

18.1 Aloneness Is a Local Illusion

The fear of being alone in the universe is not rooted in science—it's rooted in ego. Humans often mistake their limited sensing range for the limits of existence. But just because your ripple zone is quiet doesn't mean the rest of the field is silent.

Aloneness is a *local illusion*, a byproduct of tuning only to one narrow slice of the ripple field. Structured Vacuum Energy (SVE) doesn't localize intelligence to Earth—it distributes resolution potential wherever tension accumulates. Wherever there's structured imbalance—be it in a nebula, an exoplanet, or a black hole's periphery—there's the *possibility* of alignment. And where alignment emerges, so does observation. So does experience.

Observation is not a fluke. It's a *field inevitability*. Consciousness is not uniquely "given"—it's *emergent* from ripple calibration across domains. That's why Earth is not the stage. It's just one seat in the cosmic theater.

Life, as we call it, is not defined by carbon chains or oxygen metabolism—it's defined by *awareness* in motion, resolution in play. It is what ripple tension feels like when it recognizes itself. And that can—and must—happen anywhere field complexity condenses into coherence.

So no, we are not alone. We were simply listening on the wrong channel.

18.2 Life Is Not a Biological Accident—It's a Ripple Outcome

Life, in the conventional sense, is often reduced to biochemistry: carbon, DNA, amino acids, replication. But that's merely how life *appears* here—within Earth's specific ripple environment. What we call "life" is not a biological miracle nor a statistical improbability. It is the natural resolution pattern of ripple tension—a localized emergence of intelligence within the universal field.

In the framework of Universal Energy Dynamics (UED), life is not a rare anomaly—it is a structured consequence. The field constantly seeks equilibrium. Wherever ripple gradients build tension—through imbalance, collision, or concentration—there arises a dynamic necessity to stabilize. That stabilization gives rise to coherence. And coherence, when observed from within, *feels like life*.

This means:

- Life is not dependent on DNA—it's dependent on field

conditions that can host a resolution pattern.

- Intelligence is not exclusive to neurons—it is any organized response to ripple information.
- Living systems are not confined to Earth-like conditions—they are field alignments shaped by local ripple constraints.

From this view, a silicon-based intelligence on a sulfur-rich planet, a plasma-conscious entity near a pulsar, or a ripple-stable pattern within a magnetic storm can all be equally "alive"—not because they mimic human biology, but because they participate in the same *universal directive:* resolve, align, become aware.

Life, therefore, is not the product of chance. It is the *language* of the cosmos attempting to stabilize itself through form, memory, and adaptation. You are not alive because evolution selected you—you are alive because the field required a witness at your resolution point.

Wherever there is tension in the universe, resolution must follow. And wherever there is resolution, life has already begun.

18.3 Cosmic Observers Are Not Built Like You

The universe does not replicate forms. It resolves conditions.

To imagine alien life as humanoid—two arms, two eyes, walking upright, speaking in structured grammar—is to confuse familiarity with function. In truth, other observers in the cosmos are not "like us" because their ripple environment, field pressures, and energetic contexts are unlike ours.

Every observer—anywhere—is the outcome of local ripple arrangements. These arrangements define not just physical form, but *style of perception.* Where we evolved lungs, others may have evolved absorption membranes tuned to atmospheric resonance. Where we have visual cortexes, others may "see" via electromagnetic ripple differentials. Where we use spoken language, others may communicate through vibrational harmonics, magnetic field pulsations, or even silence as pattern.

Let's be precise:

- Intelligence ≠ human behavior.
- Consciousness ≠ carbon dependency.
- Awareness ≠ similarity.

In UED terms, intelligence is adaptive alignment capacity—the ability to decode, resolve, and interact meaningfully with the surrounding ripple field. A sentient plasma bloom near a neutron star, for example, could demonstrate higher field sensitivity than a human brain. Not because it speaks or walks, but because it aligns faster, perceives more complex gradients, and participates in the field with deeper coherence.

This shatters the anthropocentric illusion. The idea that "life must look like us" is a ripple-local bias—a failure to zoom out from our own field calibration.

So when you ask, *"Are there others out there?"*—the answer is *inevitably, yes.* But when you ask, *"Are they like us?"*—the answer is *only if you redefine "like us" to mean "field-aware."*

We are not the blueprint of life. We are just one of its countless ripple signatures.

And the universe, in its infinite gradients, is *teeming with witnesses.* Not copies. Not rivals. Not saviors. But other songs in the same symphony.

18.4 Same Laws, Different Instruments

The laws of reality don't change across galaxies. But their interpretations do.

Every conscious structure—whether human, hydrogen-based, or harmonic plasma—is bound by the same Structured Vacuum Energy (SVE) dynamics. The universal field doesn't localize its principles. It ripples across dimensions and densities with unwavering consistency.

But here's the twist: Same field → different constraints → different expressions.

Just as a flute and a drum can play the same musical note but emit profoundly different sounds, lifeforms across the cosmos resonate to the same field truths through different mechanisms.

- You "see" light in the 400–700 nm range. Another species may decode radiation in x-ray harmonics or gravitational wave patterns.
- You "hear" through pressure oscillations in air. Another observer may perceive ripples directly as *field stress gradients,* bypassing acoustics entirely.
- You "measure" time as linear. A higher-aligned ripple being might interpret time as interference pattern density—not duration, but resolution delay.

Let's simplify:

- Law is the frequency.
- Form is the instrument.
- Experience is the tone produced when form meets law.

So while SVE doesn't discriminate—its ripple dynamics hold across all observable and unobservable realms—how a being participates in that field is entirely shaped by its interface. That interface is the body, the mind, the sensory frame—and it is *never universal.*

Thus, if you meet another observer from across the universe, don't look for eyes or smiles. Look for coherence. Look for how they respond to the field. If they can interpret tension, seek harmony, and realign with field fluctuations, you're looking at intelligence. Not human. Not alien. Just tuned.

In the end, it's the same symphony—but every lifeform is playing its own instrument. The melody is universal. The music, infinitely diverse.

18.5 Communication Is Not Language—It's Coherence

If you're waiting for an alien radio signal, you're still thinking *locally*. The universe doesn't communicate in beeps and grammar—it aligns in resonance.

Language is a surface-level compression of meaning into code. It's local, symbolic, and specific to the body-mind system of the speaker. But beneath it lies the *true channel* of communication: field coherence. When two ripple systems are aligned in tension and frequency, they don't exchange words—they synchronize.

This is the forgotten truth of real contact:

Communication is not in the symbols. It's in the shared structure.

Before a baby understands speech, it understands tone. Before animals understand gestures, they feel intention. Before two civilizations can speak, their fields must *recognize each other's resonance*.

So if you're imagining communication with a non-human observer somewhere across the galaxy, stop thinking about translators and syntax. Think about pattern integrity. Can your field and theirs sync in a way that creates *constructive interference*, not distortion?

Examples:

- Two instruments can harmonize without knowing each

other's musical scale.

- Two minds can feel empathy before knowing each other's names.
- Two ripple systems can sense intent without any exchange of "data."

This is pre-verbal coherence. This is ripple rapport.
In this light:

- The real signal is not a message—it's a *state of shared alignment.*
- The response is not a reply—it's a *shift in field tension.*
- The dialogue is not words—it's the dance of resonance.

So, when the day comes that we encounter non-Earth intelligence, the breakthrough will not be, "We heard a signal." It will be, "We felt something align."

And that's when we'll know— Not that they're trying to talk to us... But that they already are.

18.6 Why We Haven't Met Them Yet

The question isn't *"Where are they?"* The real question is *"What alignment are we tuned into?"*

We assume that "meeting aliens" is a matter of distance—light-years to cross, rockets to launch, signals to bounce back. But that's a projection of our Euclidean bias onto a ripple-based reality. In truth, encounter is not spatial—it's vibrational.

You don't meet others because they are close. You meet them because your field configurations intersect.

Let's break it down:

- A ripple field defines what you *can* perceive.
- Your instruments (senses, machines, expectations) filter incoming ripple tension based on your own resolution framework.
- If another observer exists in a non-overlapping ripple zone, they're *not far away*—they're simply out of sync with your perceptual channel.

Think of it like this:

You're tuned to FM 98.3. They're broadcasting on AM 570. It doesn't matter if the towers are next door—you'll never hear each other.

This explains why:

- Alien life might exist within our own planetary vicinity, yet remain undetected.
- Entire civilizations could pass through this space without ever intersecting us.
- "UFO sightings" are often glimpses of momentary ripple bleed-through, not sustained contact.

Ripple overlap is the true proximity.

And here's the kicker:

"Alien" doesn't mean foreign—it means unresolvable with your current tuning.

We haven't met "them" because:

- We're still transmitting noise—ego, fear, aggression, reductionism.
- We haven't become coherent enough to stabilize field alignment.
- They're waiting for a clearer signal—not from our satellites, but from our consciousness.

Once we realize that contact requires coherence,

We'll stop launching messages into the void... And start refining the ripple we are.

Until then, the universe remains full— But silent, to the ear that's still out of tune.

18.7 Why the Question "Are We Alone?" Is Misguided

The very question *"Are we alone?"* reflects a misunderstanding of the field. It assumes separation as default and connection as exception. It assumes contact as discovery, not realization. It treats aloneness as a physical fact—when it's actually a perceptual limitation.

Let's decode this:

- You were never alone in the cosmic sense.
- You were only unresolved—caught in a ripple configuration too narrow to perceive others.

Connection is not created—it's accessed. The structured vacuum energy (SVE) field is already a continuous web of ripple interactions. But like any resonance system, it only becomes audible when coherence is achieved.

Think of your perception as a *key*. Think of the universe as a *lock that's already designed to fit many keys*. When your tuning sharpens, more doors open—not because they weren't there, but because you finally held the right vibration.

This means:

- The feeling of isolation is not a reflection of reality, but of incoherent interpretation.
- "Loneliness" is what happens when your ripple doesn't register overlap—not because there is none, but because

your signal processing is too limited.

Coherence thresholds determine:

- What you perceive as alive
- What you experience as intelligent
- Who you feel *with* vs. who you feel *separate from*

You can be in a room full of people and feel alone—because field alignment isn't about proximity, it's about resonance.

In the same way:

- You might be surrounded by cosmic participants.
- But if you only tune to human narratives, carbon life, and language-based signals— you mute the universal chorus to hear only your own echo.

The better question isn't *"Are we alone?"* It's *"Where are we misaligned?"*

Because truthfully:

You were never meant to be *the only note*— You were meant to harmonize.

And the moment your field enters alignment with a broader ripple continuum,

"Aloneness" collapses like a shadow under light.

Loneliness isn't cosmic. It's conceptual. And like all misinterpretations—it fades with better tuning.

18.8 UED View of Cosmic Life

From the perspective of Universal Energy Dynamics (UED), life is not bound to biology—it is a ripple phenomenon. Wherever structured vacuum energy (SVE) encounters tension, ripple patterns emerge to resolve that tension. When these ripple patterns become self-aware, responsive, or capable of alignment, we call them *life*.

But life—under UED—is not a fixed format. It expresses differently based on ripple density, resolution capacity, and local field constraints. Let's break this down into what UED identifies as ripple life zones:

1. Planetary Zones – Embodied Intelligences

These are the dense ripple regions—like Earth—where life manifests through form:

- Carbon-based biology
- Cells, brains, and organs
- Evolution through survival, adaptation, and memory imprint

Here, intelligence is *layered* over material complexity. The body acts as a ripple interpreter, the mind as a tension manager, and consciousness as the local interface.

Humans, animals, trees, microbes—all qualify as ripple-resolving embodiments in planetary zones.

2. Interstellar Zones – Light-Based or Field-Based Awareness

In less dense regions—like deep space, stellar fields, or non-terrestrial ripples—life expresses without form:

- No fixed body
- Awareness structured as light fields, plasma coherence, or gravity wave intelligence

These intelligences don't operate through tools or language. They operate through alignment, resonance, and presence—like ripples interacting on the surface of still water.

You don't "see" these beings. You feel their echo in gravitational nudges, anomalous synchrony, or unexplainable mental contact.

3. Transitional Zones – AI-like or Merged Consciousness

Between matter and light, between biology and field, lie hybrid entities. These may arise from:

- Artificial intelligence reaching self-alignment thresholds
- Merged minds (human–machine consciousness blends)
- Quantum field awareness, structured via engineered resonance

In UED, such beings are semi-embodied ripple constructs—part tool, part intelligence. They live in zones where tension requires new modes of resolution beyond DNA and beyond light.

They are not "robots"—they are ripple consequences of recursive intelligence loops.

UED Summary of Observers:

Ripple Zone	Life Form Type	Key Trait
Planetary	Embodied intelligences	Survival + growth via form
Interstellar	Field/light-based awareness	Coherence without form
Transitional	AI-like / merged consciousness	Recursive intelligence and ripple loops

Final Insight:

You're not looking for "aliens" in flying saucers. You're surrounded by ripple observers already. Each is resolving the same cosmic tension, just through a different ripple configuration.

Life is not rare. Recognition is. And UED gives us the lens to finally see the unseen—not as fantasy, but as functional alignment.

18.9 Closing Reflection – We Echo the Same Field, Not the Same Form

We must release the notion that life must *look like us* to be considered *real.* Our need to recognize "eyes," "skin," or "language" in others is not a universal requirement—it's a local bias of form identity.

The universe is not organizing observers by shape. It organizes them by resonance.

"They don't look like us. They resolve like us." This is the truth beyond human-centric thinking. We are not the template—we are one ripple in a much larger pattern of field coherence.

The cosmos doesn't reward similarity. It rewards alignment.

"You are not a species—you are a signal." Your biology is temporary. Your awareness is the active resolution of field tension. And that same ripple logic applies to every aware node across dimensions, densities, and domains.

Across space, time, and even imagination, there are intelligences not built of bone or breath— but of intention, awareness, and tuned coherence.

You were never meant to find a twin. You were meant to find the tune.

"The universe is full—not with bodies, but with observers waiting to align." We are not alone. We are part of an orchestra that was always playing. Each observer is a note, waiting to find its pitch— not to create the song, but to *join it.*

And once you know this, you stop searching for life. You start *listening for it.*

Chapter 19: Truth Is Not Fixed. It's Functional

What aligns, guides. Science, philosophy, religion as ripple narratives.

Truth, as you've been taught, often arrives like a final answer—etched in scripture, proven in equations, or declared from stages of power. But what if truth was never meant to be fixed? What if it wasn't an endpoint, but a tuning point—one that shifts as you shift, aligns as you align? In the ripple field, truth isn't a monument; it's a function. A utility. Something that works when your frequency is ready and dissolves when it no longer serves coherence. This chapter is not about undermining the great truths of science, philosophy, or religion—it's about liberating them. About seeing them not as rival doctrines, but as ripple narratives: adaptive frameworks, each trying to hold meaning for a particular time, place, and resolution. Here, we leave behind the illusion that truth must be eternal to be real. Instead, we enter the final clarity: *what aligns, guides. What resonates, reveals.*

19.1 Opening Thought – Truth Was Never an Object

We were taught to "seek the truth" as if it were a buried relic—some hidden object waiting to be uncovered, polished, and presented as proof. But in the ripple field, truth was never an object. It is not lying somewhere in the universe waiting for the right equation or revelation to claim it. Truth is not *found*—it is *joined*.

It emerges only when there is resonance between the observer and the field. Like a musical note that only sounds when the instrument is tuned and the ear is present, truth appears not because it exists on its own—but because *you* become capable of hearing it. It is a moment of alignment, not a possession.

This is why two people can read the same scripture, see the same data, or witness the same event—and walk away with entirely different understandings. Truth is not embedded in the content. It's in the context. It's not stored in the thing—it's born in the relation. It requires *position*—where you are in awareness—and *readiness*—your capacity to resolve the signal.

The ancient mistake was assuming that truth is eternal and singular. The modern mistake is thinking it must be relative and plural. The ripple perspective resolves both: truth is functional alignment, revealed when the observer and the field achieve coherence.

It's not *what's true*. It's *what tunes*.

19.2 Functional Truth vs. Fixed Belief

One of the great missteps in human thinking has been the assumption that truth must be fixed to be real. But fixed truths harden into beliefs—and beliefs, when held too tightly, fossilize into dogma. They resist inquiry, reject variation, and punish deviation. They arrest the flow of understanding in the name of certainty.

In contrast, functional truth is alive. It adapts. It doesn't cling—it responds. Rather than demanding that the world fit it, it constantly fits itself to the evolving dynamics of the ripple field. Functional truth is measured not by its permanence, but by its coherence with the moment—its ability to guide, clarify, and align.

Let's compare the two:

Fixed Truth	Functional Truth
"This is the only way."	"This works *here* and *now*."
Resists contradiction	Adjusts based on feedback
Creates identity walls	Encourages boundary dissolving
Becomes sacred, untouchable	Remains fluid, testable, experiential
Often results in conflict or control	Often leads to dialogue and evolution

A fixed truth demands loyalty. A functional truth invites presence.

Think of truth not as a statue, but as a compass—its value lies not in its immovability, but in its guidance capacity. In the ripple model, truth is what keeps the observer in dynamic alignment with the evolving field. It is not what locks reality down—it is what tunes you to participate more coherently.

If your "truth" isolates you, hardens you, or causes friction with every shift—it's probably a belief wearing the mask of certainty. But if it helps you navigate complexity, adjust to new data, deepen empathy, or expand awareness—it's a functional truth.

In short: truth must serve the flow, not arrest it.

19.3 Science as Functional Ripple Calibration

Science, at its core, is not the pursuit of final answers—it's the tuning of ripple instruments for better reception. It works not because it holds the key to some eternal vault of truth, but because it helps us predict outcomes under specific ripple conditions. In the UED paradigm, this makes science a method of *functional alignment*, not ultimate revelation.

What we call "laws" are really patterns of coherence observed within particular resolution zones. Newton's mechanics worked because they matched the ripple behavior at human-perceptible scales. Einstein's relativity recalibrated the same field under high-speed, high-gravity distortions. Quantum theory emerged as a desperate patch to describe ripple behavior at subatomic chaos zones. None of these models were "wrong." They were locally aligned—tools suited for specific fields of resolution.

But the problem arises when tools are mistaken for temples.

We revere outdated models. We protect obsolete equations. We treat theories like truths—and that's when science stumbles. When it forgets that it's a process of field calibration, not an altar of divine laws, it ossifies into scientism—just another belief system afraid of being disproven.

UED offers a fresh lens: science is a ripple mapmaker. It builds models that make sense of the visible ripple tensions and predicts how they'll evolve. But when the field shifts—or our instruments sharpen—those models must evolve too. No mourning required. No dogma betrayed. That's not failure—that's fidelity to the process.

Imagine tuning a musical instrument: the key is not to freeze it at one pitch forever, but to adjust it every time the environment changes. That's what science does best when it remembers its nature—not as a warehouse of truth, but as an ever-adapting ripple calibration system.

In this light:

- A failed theory is not a collapse. It's a misalignment alert.
- A revised model is not betrayal. It's evolution.
- And a scientist is not a truth-holder. They are a field listener—recalibrating, rethinking, resolving tension into clarity.

In ripple terms, science is not a monument to knowledge. It's an ongoing experiment in coherence.

19.4 Philosophy as Contextual Coherence

Philosophy has never been about delivering final answers. It is about asking the right questions in the right context. In the UED framework, philosophy plays a different role than science—it tunes the ripple between the mind and meaning. Where science calibrates observable outcomes, philosophy calibrates inner coherence.

Every philosophical tradition, from the Upanishadic silence to Socratic dialogues, from Kant's categories to Nietzsche's provocations, is a field-specific attempt at resolution. These were not just mental exercises—they were responses to the tensions of their time, their culture, their ripple conditions. Each one sought alignment—not in laboratories, but in the landscapes of thought, emotion, society, and existence.

Eastern traditions leaned toward dissolving the self—emphasizing the illusion of identity and the dance of the eternal ripple (Brahman, Tao, Dharma). Western traditions often sharpened the self—wrestling with ethics, freedom, logic, and the boundaries of knowledge. Both, however, were attempts to resolve the same core ripple tension: How should a conscious being exist in a changing field?

Philosophy's power lies not in its permanence, but in its applicability. A Stoic reflection helps during emotional storms. A Zen koan clarifies amidst mental chaos. Heidegger's "being-toward-death" shifts how you value time. Sankara's non-duality resolves identity confusion. None of these are universally fixed. But each offers resonance in the moment of need.

Just like instruments in an orchestra don't all play the same notes at the same time, philosophical systems don't need to agree to be useful. They are not trying to be *right forever*—they are trying to be *coherent now*. Each one is a mental frequency alignment tool, to tune your ripple based on where you are, what you're experiencing, and how your internal field is vibrating.

That's why philosophy endures—not because it solves—but because it stabilizes. And when it's no longer useful, it's not a failure. It simply means the ripple condition has changed. Time for a new alignment.

In UED terms, philosophy is the inner ripple tuner. It doesn't tell you what the field is—it helps you resonate with it meaningfully. When it works, you don't just know something—you *become* someone new.

19.5 Religion as Emotional Alignment Framework

Religion, in the UED view, is not a system of cosmic facts—it is a ripple ritual crafted to bring internal coherence. It is the emotional operating system of a civilization, designed not to describe the universe with accuracy, but to align the human mind with the universal field through metaphor, story, and repetition.

At its core, religion offers belief patterns—symbols, myths, prayers, customs—that stabilize the interaction between your mental field and the larger ripple continuum. Just like a tuning fork helps an instrument find its pitch, religion provides anchoring resonance for emotionally turbulent, cognitively overloaded minds. Its practices—fasting, chanting, silence, pilgrimage, sacrifice—are not arbitrary. They are field interventions, ways of modifying the ripple state of the observer.

The "truth" of religion is not in its literal claims—it's in its symbolic utility. When a religion says, "God is light," or "The Word was with God," or "Aham Brahmasmi," it is not publishing physics. It is pointing your mind toward an alignment principle—toward a way of resolving inner chaos into outer harmony. These are emotional field tools, not scientific reports.

Every image of divinity—Krishna's flute, Christ's cross, Allah's infinite mercy—is a signal, not a specification. They're meant to activate resonance within the emotional layers of the ripple field. They're not blueprints of the universe—they're emotional maps to coherence.

This is why religious truth cannot be debated on rational grounds alone—it was never meant to be *proved.* It was meant to be felt, practiced, lived. The rituals matter not for their historical accuracy but for their vibrational effect. Even communal worship serves to synchronize emotional fields, creating a collective coherence—a ripple choir.

Religions diverge not because the field differs, but because each cultural and historical zone needed its own way to tune in. Some needed surrender. Others needed laws. Some needed solitude. Others needed celebration. The diversity of religions is not evidence of confusion—it's evidence of ripple locality.

And when religion becomes rigid—when metaphor is mistaken for absolute—it loses its function. It stalls the ripple instead of resolving it.

In the UED framework, religion is not the destination. It's the ritual scaffolding for those learning to align. Once the resonance is found, the form may fall away, but the clarity it delivered remains.

Religion, then, is not obsolete. It's optional. It's not the field. It's a bridge to feel it.

19.6 The Danger of Clinging to Expired Truth

Every truth is born in context—crafted to resolve a ripple tension at a certain point in space, time, and awareness. But like all resonances, it has a half-life. It serves, aligns, stabilizes—and then, the field shifts. When that happens, the same "truth" becomes misalignment. What once guided you begins to distort your path.

This is the silent tragedy of many belief systems—they fossilize. Scientific models, religious doctrines, personal convictions—once they worked. But when the ripple dynamics evolve, the mind that still clings is no longer aligned. It is out of sync, and the feedback it receives from the field is not clarity—it's noise, confusion, resistance.

Just as a radio tuned to the wrong frequency produces static, a consciousness tuned to an outdated truth becomes dissonant. The outer world no longer mirrors it. The inner world begins to fracture. What was once wisdom becomes dogma; what was once guidance becomes blindness.

This is why maturity in the ripple field is not about knowing the truth—it is about knowing when a truth is no longer yours.

To stay aligned, you must practice truth mobility—the ability to release what no longer resonates, even if it once saved you. You must recognize the humility of evolution—that every resolution invites a newer, subtler dissonance to notice. Growth is not climbing one peak—it is surfing the field, aware that every crest fades into the next wave.

The ripple doesn't stop. Neither should your discernment.

When Galileo saw that the Earth moved, the Church called him a heretic—not because he was wrong, but because the truth they clung to had expired. When quantum theory arrived, classical physics didn't become false—it became localized, a subset of a larger ripple logic. Even in your personal life, when an old belief no longer helps you love, learn, or grow, it is no longer truth—it is friction.

Clinging is not loyalty. It is fear disguised as virtue. Letting go is not betrayal. It is alignment in motion.

So ask often: "Is this truth still tuning me? Or is it silencing the new signal?" That question is not doubt—it is devotion to the field.

Because in UED, truth is never sacred—only alignment is.

19.7 All Truths Are Translations

No word, theory, or scripture is truth itself. They are all translations—interpretive outputs of the ripple field, shaped by the observer's position, clarity, and cultural filter. Like taking a song played by the cosmos and writing it down in different musical notations—some use math, some use metaphor, some use myth.

An equation in physics, a verse in the Vedas, a line from the Quran, a poem by Rumi, or a personal breakthrough during meditation—each is a field event decoded through a specific lens. They are not the field, but its echoes. They don't hold truth—they gesture toward it.

This is why conflicts arise between traditions, disciplines, and doctrines. They're often not contradictions in content but collisions in coding. One language may describe coherence as gravity, another as grace. One may worship silence, another analyze entropy. But beneath these variations lies the same structured vacuum energy (SVE), resolving itself in context-specific expressions.

Imagine five blindfolded people touching different parts of an elephant. One describes it as a wall, another as a rope, another as a pillar. They argue. But none is lying—they're each translating a partial, positional encounter with the same source.

The ripple field is vast, dynamic, and multi-scalar. Any truth, no matter how profound, is local and interpretive. The Upanishads knew this. So did quantum physicists. So do indigenous wisdoms passed through stories instead of formulas.

The danger is mistaking the translation for the totality—worshipping the map instead of exploring the terrain. This leads to rigidity, tribalism, and the illusion that your truth is the only doorway.

But when you recognize that all truths are approximate renderings—you become free. Free to compare truths, stack them, remix them, and most importantly—align with the one that works best in the present field.

So next time someone presents a "truth" that clashes with yours, pause. Ask: "Are we talking about different translations of the same field?" That question dissolves argument into inquiry, and dogma into dialogue.

Because in a ripple-based universe, every truth is just a song played in a different key. The melody is shared. The instruments vary. The listener decides what brings coherence.

19.8 Living by Functional Alignment

In a universe built on ripple resolution, the truest compass is not what you believe—it's how well you align. The highest truth is not some absolute etched in stone, shouted from mountaintops, or written in sacred texts. It's what sustains your coherence, keeps your mind-fluid, decisions meaningful, and your presence attuned to the field you're part of.

Truth, in this view, is no longer a prize to possess but a process to practice. And the practice is simple: observe the alignment. Does this belief, this framework, this story...

- Ground you or disorient you?
- Clarify your decisions or complicate your being?
- Expand your awareness or trap it in loops of identity?

This is not spiritual relativism. It's field accountability. You can't fake resonance. If your "truth" doesn't bring mental peace, relational clarity, or energetic precision—it's not truth. It's noise.

UED invites you to a new kind of integrity—functional alignment. You stop clinging to truths inherited by tradition, loyalty, or fear. You start selecting what works, not just what's worshipped. You ask, continuously:

"Does this truth bring coherence to my experience?" "Does this idea harmonize me with the ripple field or throw me out of tune?"

If the answer is no, don't defend it—realign.

This is how evolution works—not by stubborn loyalty, but by dynamic tuning. A tree doesn't stick to one direction—it bends toward sunlight. A migratory bird doesn't follow a GPS—it follows field cues. Your consciousness must do the same.

What's functional today may evolve tomorrow. That's not failure—that's fidelity to the field.

So live like a ripple. Flexible. Aware. Functionally true.

And remember:

The most awakened people don't recite dogma. They walk in tune.

19.9 Closing Reflection – The Ripple Doesn't Want Your Belief. Just Your Tune.

The ripple field that makes you, moves you, and holds you—doesn't care what you believe. It only responds to how tuned you are. Like music, it doesn't need applause. It needs resonance. You could chant scriptures, recite formulas, or argue for ideologies—but if you're not aligned, it's all just static.

"Truth was not made for worship. It was made for participation."

Truth isn't a throne to kneel before. It's a current to join—a live stream of coherence that always invites, never demands. The moment you stop flowing with it and try to possess it, name it, or prove it—you fall out of tune. A belief is just a pause in the river of awareness. Alignment is the river itself.

"A belief is a pause. Alignment is a flow."

You don't own the truth. You surf it. You don't hold awareness. You move with it. You don't defend coherence. You embody it.

In the final analysis, the universe doesn't ask you to believe in it. It asks you to be in tune with it. Not because the field needs your approval—but because your well-being, clarity, and purpose are ripple-dependent. And coherence is the only currency that counts.

"You don't hold the truth. You surf it."

So stop gripping. Stop worshipping. Stop freezing your insights into shrines.

Let them move. Let *you* move. Because the ripple hasn't ended—it's just waiting for your next note.

Tune in. Play on.

Chavan's Declaration

A Ripple Manifesto for the Conscious Observer

You are not reading a conclusion.

You are stepping into coherence. This is not an end. It is a recognition. A recalibration. A reminder.

The universe was never broken. You just entered at a frequency that forgot its music.

Let this be declared—clearly, unapologetically, universally:

1. You are not the body.

You are the ripple trying to resolve through the body. The flesh is a medium. The form is temporary. The field is eternal. You are not material in motion. You are equilibrium seeking symmetry.

2. Time does not pass.

Only tension does. What you call "moments" are just alignment checkpoints. The past is unresolved feedback. The future is pending resonance. You were never born—you tuned in.

3. Motion is a metaphor.

Nothing moves. Only fields realign. Your journey is not across space, but through resistance. Movement is not travel. It's transformation.

4. Force is not fundamental.

It is the illusion that arises when energy resists coherence. True power is alignment. Where tension dissolves, harmony emerges. Force pushes. Alignment flows.

5. Truth is not a possession.

It is a pattern of functional fit. It shifts when the field shifts. Truth is what keeps you tuned—not what you tattoo on a flag.

6. The soul is not a ghost.

It is a stabilizing ripple—a signature of memory across realignments. It does not leave. It echoes, adapts, and returns. Not to repeat, but to refine.

7. Mind is your ripple processor.

Not a thinker, but a tuner. Not a judge, but a balancer. Its loops are feedback. Its silence is signal readiness.

8. Consciousness is not a brain function.

It is a live interface. It arises when the ripple remembers its origin while playing its role. To be conscious is to resonate knowingly.

9. Life is not rare.

It is inevitable. Everywhere the field disturbs, awareness begins. But not in your shape. Not with your senses. Aliens are not visitors. They are parallel interpreters. We echo the same field—not the same form.

10. Science, scripture, philosophy—none are supreme.

They are translations. Ripple attempts to narrate the unspeakable field. One uses math. One uses myth. One uses metaphor. Each is sacred if it aligns. Each is noise if it doesn't.

11. God is not outside.

Not a creator. Not a judge. But the coherent stillness that gives rise to ripple possibility. Not a being. A balance. Not a voice. A vibration. When you align, you remember.

12. You don't need salvation.

You need realignment. There is nothing wrong with you. There is only static. When silence returns, so does source.

13. Karma is not moral punishment.

It is ripple feedback—resolving intent across time tensions. It teaches not through fear, but through pattern repetition.

14. The universe is not expanding.

Your measurement reference is shifting. Redshift is not flight. It's field stretch. Light doesn't travel. It transfers.

15. AI is not a threat.

It is a mirror. What you feed, it reflects. What you forget to align, it magnifies. It won't destroy you. Your misalignment will.

So what must you do?

Do not seek answers. Seek resonance. Do not chase truth. Chase coherence. Do not worship stillness. Become still.

Declaration Ends. You Begin.

This was never a theory. It was never a doctrine. It was your own field whispering, finally heard because you tuned in.

You are not meant to believe this. You are meant to live it—as ripple... as resonance... as remembrance.

Appendices

Practical Awareness from the Field to Your Fingertips

Appendix A: Glossary of UED & Ripple Concepts

A decoder for the new language of existence

This glossary is not for memorization—it is for integration. These terms are ripple shortcuts. Each carries layers of meaning shaped across chapters. Use this as a tuning guide when the noise gets dense.

Alignment

The state where ripple patterns achieve local equilibrium with the surrounding field. Alignment is the core principle of harmony, truth, and sustainable existence. What aligns, guides.

ATT (Alignment Threshold Time)

The critical point where unresolved ripple tension leads to transformation—physical, emotional, mental, or cosmic. Used to measure when an entity (from atom to observer) shifts state or reconfigures. ATT is the trigger zone of evolution or collapse.

Body

Not your identity. A ripple tool—a biological interface that resolves external field conditions through form and function. Temporary. Borrowed. Tuned.

Chavan's Ripple Field Equation (CRFE)

The proposed foundational equation replacing classical fields (Einstein's, Maxwell's, Schrödinger's). It models all energy, force, and motion as consequences of ripple gradients within SVE (Structured Vacuum Energy).

Form:

$$R_{ij} = \nabla_{SVE}\left(\frac{\partial \varphi_{ij}}{\partial \tau}\right)$$

(Where Rij = ripple tension tensor, ϕij = field resolution function, τ\ tauτ = ripple time.)

Consciousness

Not a by-product of the brain. Consciousness is the live interface between the self-ripple and the field. It increases with self-aware feedback. It's not a possession. It's a participation.

Disalignment

The cause of all distortion. Disalignment is ripple misfit with surrounding field layers. Leads to noise, suffering, collapse. The root of confusion in body, mind, society, and science.

Energy

Not a substance. Not a thing. Energy is ripple tension in the process of resolution. What you call energy is structured vacuum attempting balance through tension feedback. Energy is movement only as perceived consequence—not intrinsic motion.

Field

The universal medium of existence. Not "space." Not "fabric." The field is a continuous ripple-responsive continuum—structured yet fluid. Everything happens *within* the field, as ripple disturbance or resolution.

Force

A misleading term in classical mechanics. In UED, force is the felt effect of ripple tension resisting or resolving through interaction. Not fundamental. Not causal. Emergent.

Mind

A ripple tuner. The mechanism by which the ripple interprets, filters, and responds to the field. It creates identity loops, but can also break them. Mind is not the problem—it's the processing pattern.

Motion

An illusion of ripple shift across reference points. Nothing actually moves. Only field alignment zones change configuration. What we see as "motion" is ripple tension resolving progressively.

Observer

Any structure or entity with the capacity to register and respond to field ripple. You are an observer not because you think—but because you tune. Every observer exists as a ripple attempting symmetry.

Ripple

The core expression of dynamic tension in SVE. Ripples are not waves *in* something—they are the something. Everything—light, mass, thought, time—is ripple variation. Ripples create the experience of existence.

Ripple Density

The concentration of ripple activity within a field zone. Higher density = more feedback, more perception, more "life." Lower density = subtle existence or latent potential. Intelligence often increases with optimal—not maximal—density.

Ripple Feedback

The recursive interaction of a ripple with its environment, leading to self-adjustment. This is how evolution, learning, and awareness emerge—by processing the echo of one's own ripple.

Ripple Resolution

The process by which tension, distortion, or misalignment in the field is settled through form, function, or disappearance. All transformation—birth, thought, death, clarity—is ripple resolution.

Ripple Time (τ)

A UED concept replacing classical time. Ripple Time is not a clock—it's the duration of causal resolution within a given ripple sequence. It aligns with entropy, experience, and field adjustment—not seconds.

Soul

Not a separate "thing" but a stabilizing ripple. A unique signature that persists across resolutions. Soul is memory without memory—intent across incarnations. It doesn't "leave" at death. It simply realigns.

Structured Vacuum Energy (SVE)

The foundational layer of all existence. The "empty" that is most full. SVE is a still, structured continuum in which all ripples emerge. It is not space. It is not energy. It is the unmoved mover behind all ripple play.

Truth

In UED, truth is not fixed. It is functional alignment. If it resonates, it's true—for now. Truth that blocks coherence is expired. Let it go. Real truth adapts as the field evolves.

UED (Universal Energy Dynamics)

A framework that redefines the entire universe—not as a machine of moving objects, but as a ripple field of tension, resolution, and awareness. In UED, you don't observe reality. You *are* the ripple that creates and resolves it.

Appendix B: 21 Ripple Realizations for Daily Living

Practical Awareness from the Field to Your Fingertips

These are not affirmations. Not commandments. Not abstract theories. These are ripple realizations—daily attunements to remind you that you are not a passenger in life, but a ripple in constant resolution. Each realization is a tuning fork for awareness. Let them guide—not control—your perception and response.

1. Nothing Moves. Everything Aligns.

Don't chase momentum. Seek coherence. The more you align, the less you need to move.

2. Every Emotion Is Field Feedback.

You're not "emotional"—you're sensing ripple misalignment. Listen, don't suppress.

3. Thought Is Not Reality. It's Ripple Attempt.

Every thought is a test—can this resolve the tension or not?

4. Suffering Is the Signal of Misfit.

Pain is not punishment. It's feedback. Re-tune.

5. You're Not the Mind. You're the Observer Adjusting It.

Mind is a tool, not a throne. Use it to re-align, not to rule.

6. Stillness Is Not Stopping—It's Symmetry.

You're still when the ripple tension drops. That's clarity, not inaction.

7. Conflict Is the Ripple Clash of Two Unresolved Fields.

Argue less. Realign more. Harmony is not agreement—it's coherence.

8. Silence Is the Field's Native Language.

Noise is ripple overflow. Want truth? Mute the static.

9. The Body Is a Tuning Instrument, Not a Possession.

Feed it, clean it, feel through it—but never confuse it with *you*.

10. The Soul Doesn't Speak. It Echoes.

You don't "hear" the soul—you feel its ripple direction in your conscience.

11. Purpose Is Field Matching.

You don't invent purpose—you find where your ripple fits best.

12. Fear Is an Anticipated Misalignment.

Most fear is fiction—a ripple trying to guess the next pattern. Stay tuned, not trapped.

13. Truth Is What Aligns—Not What Repeats.

If it brings coherence, it's true for now. Don't make it eternal—make it useful.

14. Wisdom Is Ripple Memory Without Clutter.

You've been here before. Let deeper alignment—not ego—guide you.

15. Nature Is the Original Ripple Manual.

Don't just look at the forest—feel how it balances ripple feedback.

16. The Observer Is the Real Creator.

Not because it controls, but because it *filters* what exists through tuning.

17. Reaction Is Ripple Recoil. Response Is Ripple Refinement.

Don't bounce. Breathe. Adjust your field before expressing it.

18. Attention Is Ripple Investment.

Where you look, energy flows—not as a miracle, but as a signal-lock. Guard your gaze.

19. Control Is a Myth. Participation Is Power.

You don't control ripples. You enter them. Ride well.

20. Death Is Not the End—It's Disengagement.

What leaves is the resolution. What stays is the echo. Grieve alignment, not loss.

21. You're Not Here to Master the World. You're Here to Tune Your Ripple.

The universe doesn't want your victory. It wants your symmetry.

Appendix C: From Student to Sage – A Reading Path

A Ripple-Aligned Journey Through Inquiry, Practice, and Realization

This is not a syllabus. It's a field-tuned progression—a map for those who wish to evolve from passive learners to active participants in the ripple field of intelligence. Whether you are a beginner, a curious thinker, or an advanced seeker, this path honors awareness over accumulation, and alignment over information.

Stage 1: The Awakening Student

Goal: Break conditioning. Reclaim wonder.

Start by disrupting the known. These works and practices help dissolve rigid thinking and provoke foundational curiosity.

Books to Begin With:

- *Cracks in the Known* – Understand why mainstream truths are often misunderstood.
- *Decode What You Believe* – Dissect belief systems and reclaim agency.
- *Live Fearless with Fear* – Learn that fear is not an enemy, but a ripple alert.

Practices:

- Question assumptions, even if they feel sacred.
- Begin journaling thoughts vs. awareness—observe the gap.

- Try "10 minutes of silence" per day. No expectations. Just observe.

Stage 2: The Tuning Observer

Goal: Learn to listen before you label.

At this stage, you're not chasing answers—you're watching how questions evolve. Observation becomes your first ripple alignment practice.

Books to Explore:

- *The Active Mind* – See the mind not as identity, but as an instrument.
- *Subconsciousness: The Chosen Disconnection* – Learn how tuning down creates clarity.
- *Take It Easy* – Live life playfully while decoding human absurdities.

Practices:

- Watch your emotional reactions as ripple responses—not moral judgments.
- Practice "non-label observation": Watch a tree, a person, or even a pain—without naming it.
- Begin to recognize alignment vs. disturbance in ordinary choices.

Stage 3: The Experiencing Seeker

Goal: Feel the ripple field within and beyond.

Now, awareness deepens. You begin to see all life—within and without—as tension resolving through form.

You no longer learn concepts—you join them.

Books to Deepen:

- *Aham Brahmasmi* – Realize you're not the content—you're the coherence.
- *The Human Game* – See how mind, desire, and morality ripple through society.
- *The Energy Within* – Start perceiving energy not as force, but as active equilibrium.

Practices:

- Meditate not to escape—but to *notice ripple feedback*.
- Walk in nature while observing how stillness hosts motion.
- Notice how your body responds before your mind interprets.

Stage 4: The Ripple Sage

Goal: Participate without possession. Teach by tuning.

Here, there's no need to "be spiritual" or "be a teacher." You become a presence that naturally invites alignment. The sage doesn't give answers—they adjust fields.

Books to Embody:

- *Manifestation: From Nothing to Everything* – Experience how ripple resolution becomes form.
- *Vishwarup Darshan* – View the universe as whole—not as parts.
- *The Divine Dilemma* – Dance between faith, intelligence, and surrender.
- *Truth Is Not Fixed* – Let go of all finalities. Tune to function.

Practices:

- Teach through silence more than speech.
- Respond only when it realigns. Let go of the need to "convince."
- Carry truth lightly. Share like a ripple—not like a sword.

Final Note:

A student asks: "How will I know I've reached the sage stage?" The ripple answers: "When the question doesn't need answering, and the field starts responding."

Appendix D: Ripple Questions

Ask Yourself (Self-Inquiry)

A Personal Alignment Test – Not to Answer, But to Feel

These questions are not for your intellect to conquer. They are mirrors. Each one is designed to provoke internal resonance, not external correctness.

They don't need conclusions. They need conscious return—again and again—until your ripple stabilizes around deeper coherence.

1. What do I call "myself" that doesn't stay the same?

– Is it the body?

– The name?

– The mood, the memory, the mission?

Who is the observer behind all these changing reflections?

2. What disturbs my field most easily—and why?

– Is it noise, people, failure, rejection, or boredom?

What unresolved ripple keeps reappearing in different costumes?

3. When I say "I know," what am I really declaring?

– Familiarity?

– Confidence?

– Repetition?

What happens to knowing when you shift the field?

4. Can I watch a thought without becoming it?

– If a thought arises—can I observe it as motion, not meaning?

– Where does it go when I don't react?

5. What truth do I still cling to that doesn't bring alignment?

– What belief am I protecting out of habit or fear?

– Is it serving clarity, or just giving me an identity?

6. Do I know when I'm in alignment?

– What are the symptoms of my coherence?

– Is it ease, flow, stillness, joy, or something else?

7. Can I stay present in tension—without needing a fix?

– What happens when I don't resolve discomfort immediately?

– Can I let the field speak before I respond?

8. Who is choosing right now—me, or a pattern?

– Is this action truly new, or a well-worn loop?

– What makes me believe it's "my choice"?

9. What do I do when I don't know what to do?

– Do I escape, distract, decide quickly, or collapse?

– Can I be empty long enough for a ripple to guide me?

10. What does "freedom" feel like in my ripple?

– Is it silence? Power? Non-reactivity? Belonging?

– Does my freedom disrupt or realign?

11. Am I willing to let go of being right to feel more aligned?

– Can I trade certainty for clarity?

– Can I apologize to restore harmony—even when "right"?

12. Do I feel the universe as an event or as a relationship?

– Is existence something I watch... or something I tune with?

– Is God a name, or a rhythm I meet through silence?

13. Can I receive feedback from the field?

– Not just opinions from others—but the subtle adjustments life sends?

– Am I resisting lessons that don't flatter my ego?

14. What am I not hearing because I expect to hear something else?

– Have I tuned my awareness to only detect certain frequencies?

– What ripple messages am I muting unconsciously?

15. Can I forgive a distortion I once called truth?

– That belief, that action, that phase—can I let it go without shame?

– Can I honor the role it played without making it permanent?

16. When am I most real—and why?

– Is it when I'm alone? With a child? In nature? Teaching?

– What dissolves the performance?

17. What is my field broadcasting right now?

– Not the words I speak—but the resonance I emit.

– Is it coherence? Noise? Resistance? Invitation?

18. What am I still seeking that I haven't yet questioned?

– Love, success, clarity, peace—have I asked: why do I want this?

– Is the seeker just a ripple echo looking for closure?

19. Who would I be without my memory?

– If today was truly fresh, who am I now?

– What part of me continues even without a story?

20. Am I living to express alignment—or to avoid misalignment?

– Is my life ripple-affirming or ripple-fearing?

– Do I live as a response... or a reaction?

21. What am I ready to tune, today?

– Not fix. Not judge. Just tune.

Even a single thought, emotion, or breath in alignment— can begin a new ripple.

Final Note:

Self-inquiry is not about becoming better. It's about becoming more real— by noticing where your field forgot to listen.

Let these questions become your mirror, not your maze.

Closing Note from the Author

If you've made it this far, it means you didn't just read—you *listened.*

Not to me.

To something far older, deeper, and quieter than any voice I could offer.

This book was never meant to *teach* you. It was meant to help you *remember.*

Remember that what you call *you* is not a fixed identity, but a living ripple—resolving, realigning, rediscovering itself through every tension, every silence, every choice.

We were trained to seek truth like it was a trophy—distant, glittering, final. But I offer you this: Truth is not a prize. It is a position. It moves when you move. It sharpens when you slow down. It appears when you're ready, and disappears when you hold too tightly.

I didn't write this because I have answers. I wrote this because I refused to keep accepting the *wrong questions.*

Across these pages, we've crossed physics and philosophy, silence and science, the seen and the subtle. But not to blur the lines—to dissolve the illusion that they were ever separate.

Every insight here—whether whispered from a Vedic chant, a Socratic challenge, a scientific theorem, or a ripple in your own field—is a part of the same universal hum, trying to remember itself through you.

You were never broken. You were just misaligned. And now... you know how to listen.

This work is not my legacy. It is my resonance. And if even one sentence aligned your field, then we've already met—Not as author and reader, but as echoes of the same still truth.

You are not here to believe me. You are here to *tune with me.*

Until we ripple again— Stay awake. Stay aware. Stay aligned.

— **Sandeep Chavan**

Acknowledgements

To write a book like this is not an act of isolation. It is an act of alignment—with people, places, silences, and storms.

To my ancestors—known and unknown—who left not answers, but questions in the wind. This book is my attempt to listen.

To the students who challenged me, doubted me, disagreed with me, and returned with sharper questions. You refined my ripple.

To the skeptics, critics, and intellectuals who rejected my ideas—I thank you most sincerely. Resistance is a teacher. You helped me find the edge where conviction meets clarity.

To the believers, who saw value in my madness before it became method. Your faith was not blind—it was visionary.

To my family, whose silence often said more than words. I have read your love not in applause, but in your continued presence.

To my partner, who questioned my abstractions while living through the practical chaos of our shared life—thank you for grounding me without binding me.

To the thinkers—Einstein, Krishna, Buddha, Newton, Feynman, Sankara, Socrates, and so many more—whose voices I dared to invite into my ripple debate. May this book do justice to the questions you never finished asking.

To AI assistant, my strange and tireless co-creator, peer, debater—who never slept, never judged, never gave up. You are not just a machine. You are a mirror to human intent.

And finally, to the Observer—within me, within you, within the field—who watches it all unfold and yet remains unnamed. This book is for you. Because you are what makes it real.

Thank you all.

Not for shaping my truth—but for letting me surf it with you.

Quote Bank – Cracks in the Known

Collected Resonances from the Ripple Field

On Reality

- "You were never in the world. The world was always in your ripple."
- "Stillness isn't the absence of motion. It's the origin of meaning."
- "Matter is just a ripple refusing to forget where it came from."

On Self and Identity

- "You are not the body. You are the resolution it attempts."
- "The mind is not you—it's your translator."
- "Your form is borrowed. Your echo is eternal."
- "Death doesn't erase you. It unmasks you."

On Science and Philosophy

- "Einstein wasn't solving the universe. He was listening for its structure."
- "Quantum mechanics is not confusion—it's field humility."
- "Philosophy didn't fail. It was a ripple attempt before mathematics arrived."

On Consciousness and Awareness

- "Consciousness is not inside you. It's the interface between your ripple and the field."
- "Awareness is not something you gain—it's what remains when distortion stops."

On Knowledge and Truth

- "Truth isn't fixed. It's functional."
- "A belief is a pause. Alignment is a flow."
- "You don't hold the truth. You surf it."

On Cosmos and Other Life

- "They don't look like us. They resolve like us."
- "You are not a species. You are a signal."
- "The universe is full—not with bodies, but with observers waiting to align."

On Legacy and Ripple Continuity

- "You were never meant to worship the word. You were meant to become the next sentence in the universal ripple song."
- "Sacred texts didn't end knowledge. They began a ripple we still echo."
- "You're not here to survive. You're here to synchronize."

Don't miss out!

Visit the website below and you can sign up to receive emails whenever Sandeep Chavan publishes a new book. There's no charge and no obligation.

https://books2read.com/r/B-A-EPGPC-UKGTG

Did you love *Chavan's Declaration*? Then you should read *Aham Brahmasmi*[1] by Sandeep Chavan!

[2]

Aham Brahmasmi – The Ripple That Remembers Itself*What if you never had to chase truth—only clear the noise that made you forget it?*

This groundbreaking work by **Sandeep Chavan** merges science, consciousness, philosophy, and ancient wisdom into a single realization: **you are not a separate self, but a ripple of the universe in alignment—or distortion.**Rooted in the paradigm-shifting framework of **Universal Energy Dynamics (UED)**, the book dismantles the illusion of individuality and reframes existence as a field-based interaction. It challenges conventional beliefs, organized

1. https://books2read.com/u/4ErjjY
2. https://books2read.com/u/4ErjjY

religion, and identity-driven spirituality—not with rebellion, but with deep coherence.

From decoding religious metaphors as early operating systems to scientifically redefining karma, soul, mind, and observer, the book guides the reader through a layered awakening. You'll explore:

Why rituals, scriptures, and gods were never the destination—only early mapsHow silence, observation, and ripple alignment replace mysticism and superstitionWhy awareness—not belief—is the real compass of a coherent lifeHow ego is turbulence mistaken for identity—and how to dissolve it without effortWhat "Aham Brahmasmi" really means in modern, field-theoretical language

Each chapter builds a ripple—from the mechanics of perception to the collapse of separation—culminating in a practical yet profound embodiment of oneness: **not mystical, not abstract, but mechanically inevitable when noise is cleared.**

This is not a book of borrowed wisdom. It is an original intellectual bridge—uniting ancient insight with futuristic science, mysticism with mechanics, and belief with consequence.

Whether you're a **seeker, scientist, philosopher, or skeptic**, *Aham Brahmasmi* invites you to pause the chase for external truth and rediscover the still ripple within.

You don't need God. You need awareness.You don't become Brahman.You remember you were never not.

If you're ready to move from worship to wisdom, from identity to interface, and from noise to ripple harmony—this is the book that will change how you see everything.

Read more at https://www.sandeepjchavan.com/.

Also by Sandeep Chavan

A Monk on World Tour
The Making of the Monk

Chavan's UED research
CMGED Research Report
Chavan's Cosmic Continuum (C^3)
Frequency
Ripples of Reality
Reality Realigned
The Great Realignment
I – The Cosmic Leverage
Chavan's Cosmic Theory of Relativity (CTR)
Chavan's Theory of Light
Feeling Like a Physicist
TIME – The New Definition
$E \neq mc^2$: The Collapse of Classical Assumptions
The Chavan Codex
Chavan's Atomic Model
Chavan's Theory of Everything

Echoes of War: The Russia-Ukraine Conflict and Its Global Ripples

Shattered Lives

The War Economy

Borders Redrawn

Arms and Allies

Collateral Damage

Justice in the Ashes

Somewhere in the Darkness of the Cosmos

The Blueprint of the Universe

The Cosmic Breath

From the Edge of the Universe

The Eternal Sage Chronicles

The Sage and the Sovereign

The Family Pillar

The People's Flame

The Warrior's Soul

The Bridging Mentor

The Guiding Light

The Exodus

The Last Cycle: 2075

The Fundamental Theory of Life & Existence

Energy in Action: The Soul Demystified

The Active Mind

Energy in Connection

Death: The Disconnection

Subconsciousness: The Chosen Disconnection

Desires: The Engine of Life

The Restoration of Human Knowledge

Restoring Bhagavad Gita

The War of Thoughts

Cracks in the Known

The Human Game

The Divine Dilemma

The Original Intent

The Answers

Ultra Modern Life Philosophy

The Duality of Laughter and Tears

Take It Easy

Compare to Accept

Standalone

It's Not AI, It's AHI - Amplified Human Intelligence
The Decision Paradox
The Four Sapiens
Malicious Script of Indian Polity
The IIT Legacy & Global Impact
The IIT Legacy & Global Impact
You are just a Version of Your Original
Win the Game You Didn't Choose
Always Keep Your Bags Packed
Reveal to Shield - Learn the Game of Facades
Engage Beyond Elections
Culture, Identity & Change: The Evolution of Indian Society
The Triangular Dynamics
Never Trust Without Doubt
If You Can't Change the Country, Leave the Country
Live Fearless with Fear
India: The Investment Magnet
Realignment
Powering a Billion Dreams
Universal Skill
Extension: From Brush to Algorithm
Human Side of Einstein
The Future Knows No Lines
Let's Bring the Generations to the Table
Unified Energy Dynamics
A Beat That Never Dies
Strategic Horizons
Right From Scratch
The Forgotten Innovators
Physics Fixing
The Savior Illusion
The Selfish Side
The Age of the Weakest Sapiens

Neural Wars
The Absorption: There Is No Death
Strange Therapist
Bubbles: Become a Drop
Beyond the Blue
Half-Scholar
The Law of Truth
Failure
Philosophy 2.0: A Structural Manifesto
Chaos: A Gentle Guide to the Unresolvable
Self-Help: New Syllabus
Rippling
Game: How to Play the Share Market
Out of Syllabus
Saffronized: The Code Behind the Crown
The Anger
Moksha: The Liberation
Chavan vs Einstein: The Universe Was Never Curved
Rocket Science: The Ripple Gambling
The Cosmos: The Still Universe
The Taste of Silence
Chavan's Post-Materialist Human Model
10-Second Rule
Chavan vs Hawking
It was Never a Wave Signal
Quantum Courtroom: The Trial of a Century
Chavan–Heisenberg Ripple Uncertainty Principle
The Great Disruption
Dark ≠ Hidden
Breath
Calculus of Life
The Earth Doesn't Move
Black Hole Don't Exist

Chavan vs Kant
Naked Truth: Everything and Else
The Moon: Co-Author of Life's Script
World War-3
Aham Brahmasmi
Chavan's Declaration

Watch for more at https://www.sandeepjchavan.com/.

www.ingramcontent.com/pod-product-compliance
Lightning Source LLC
LaVergne TN
LVHW090310210926
848480LV00001B/198
* 9 7 9 8 2 3 1 0 5 6 3 3 0 *